Shop Accessories
You Can Build

Shop Accessories
You Can Build

The Best Of Fine WoodWorking

Cover photo by Alec Waters

Taunton
BOOKS & VIDEOS
for fellow enthusiasts

First printing: 1996
Printed in the United States of America

A Fine Woodworking Book

Fine Woodworking® is a trademark of The Taunton Press, Inc., registered in the U.S. Patent and Trademark Office.

The Taunton Press, Inc.
63 South Main Street
P.O. Box 5506
Newtown, Connecticut 06470-5506

Library of Congress Cataloging-in-Publication Data

Shop accessories you can build.
p. cm. — (The Best of fine woodworking)
"A Fine woodworking book" — T.p. verso.
Includes index.
ISBN 1-56158-118-6 (pbk.)
1. Woodworking tools—Design and construction.
2. Woodwork—Equipment and supplies—Design and construction. 3. Workshops—Equipment and supplies—Design and construction. I. Fine woodworking. II. Series.
TT186.S56 1996
684'.08'028—dc20 95-41547
CIP

Contents

Introduction

Strange as it may sound, a craftsman might very well leave some of his best work in a dusty corner of the shop for months at a time, unseen by anyone else, and feel he hadn't wasted a moment of his time. That's because some of his most inventive work might well be the shop accessories he's made as part of making something else. Whether it's a tablesaw fence that stays square or a beautifully detailed toolbox, a well-designed shop accessory is a real friend. A good one might do anything from improving accuracy to making the shop a safer place to work.

In this collection of 38 articles from *Fine Woodworking* magazine, authors share some of the accessories they've built along the way. Some of them, like the panel-storage system devised by Skip Lauderbaugh, might become permanent fixtures in your shop. Others take care of specific, short-term needs. Whether your interest is in building cabinets or carving cabriole legs, you'll find something here that will make your work in the shop easier and more productive.

—Scott Gibson, editor

The "Best of *Fine Woodworking*" series spans over ten years of *Fine Woodworking* magazine. There is no duplication between these books and the popular *"Fine Woodworking* on..." series. A footnote with each article gives the date of first publication; product availability, suppliers' addresses and prices may have changed since then.

Shop-Built Roller Extension Table

Roller balls and vertical adjustability help this unit handle sheet goods with ease

by Bob Gabor

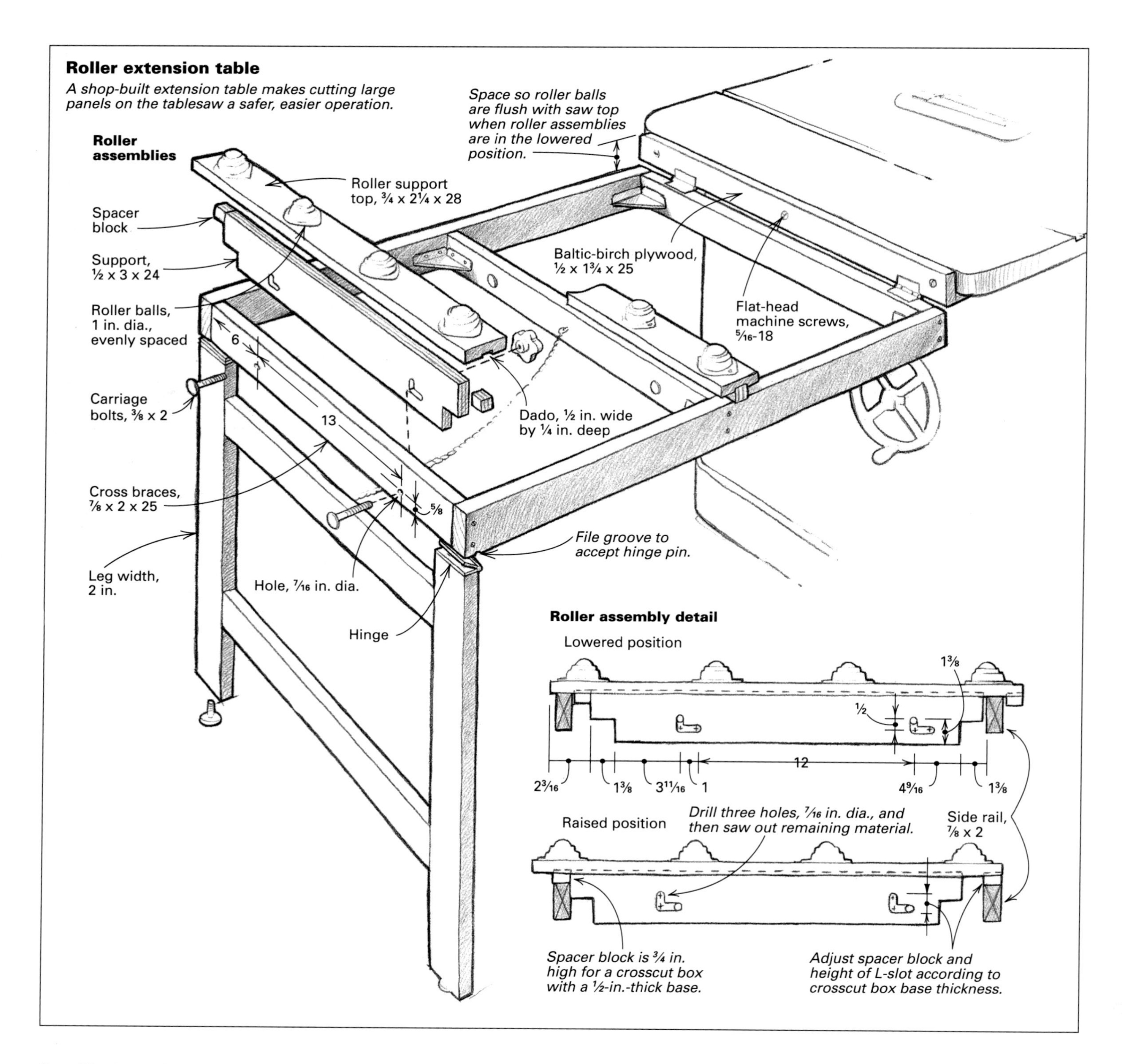

Drawing: Bob La Pointe

I was tired of wrestling big sheets of plywood across the top of my tablesaw. I already had an outfeed table on the back of the saw, but what I really needed was a side extension table to support the heavy panels going into the saw. I didn't want to give up too much valuable floor space to an accessory that I wouldn't be using most of the time.

My solution was a fold-away extension table that uses rows of roller balls to support the workpiece. I chose roller balls instead of long, tube rollers because the balls won't pull stock off-line as it is fed through the saw. Normally, the roller balls are even with the saw's tabletop, but they also can be raised to support long panels that overhang the end of my crosscut box. This straightforward shop fixture is easy to build and use. It sets up and drops back out of the way in a matter of seconds, and it makes cutting plywood on the tablesaw safer and more manageable.

The extension table drops into its stored position in seconds *and takes up no floor space. Adjustable roller assemblies can be raised so that the table also works with a sliding crosscut box.*

Utility and economy in a shop tool

I'd rather make furniture than shop tools, so I designed the extension table to be as simple as possible. The top frame and the leg assemblies, as shown in the drawing on the facing page, are inexpensive and easy to assemble with a biscuit joiner. Yet they're light and strong. The length of the top-frame assembly and the leg assembly is determined by the distance between the floor and the top of the saw.

The top frame needs to be sized to just clear the floor in the folded position. The legs must be long enough to make the roller balls level with the saw top when the frame is in the raised position.

The extension table also supports long stock in my sliding crosscut box because the rollers are adjustable by the thickness of the crosscut box's bottom. Mounting the rollers on T-shaped assemblies, which adjust easily after loosening a few knobs, was a simple and reliable solution.

To fold the unit for storage (see the top photo), I hinged the legs to the top frame and also hinged the top frame to the tablesaw top. When folded down, the table doesn't take up much room in my shop. By adding adjustable levelers to the leg assemblies, I made it easy to fine-tune the height.

Finally, I added a piece of lightweight chain to limit the leg travel and a screen-door hook to keep the leg assembly folded for storage. I've been so pleased with the roller extension table that I've built another and attached it to the side of my outfeed table. □

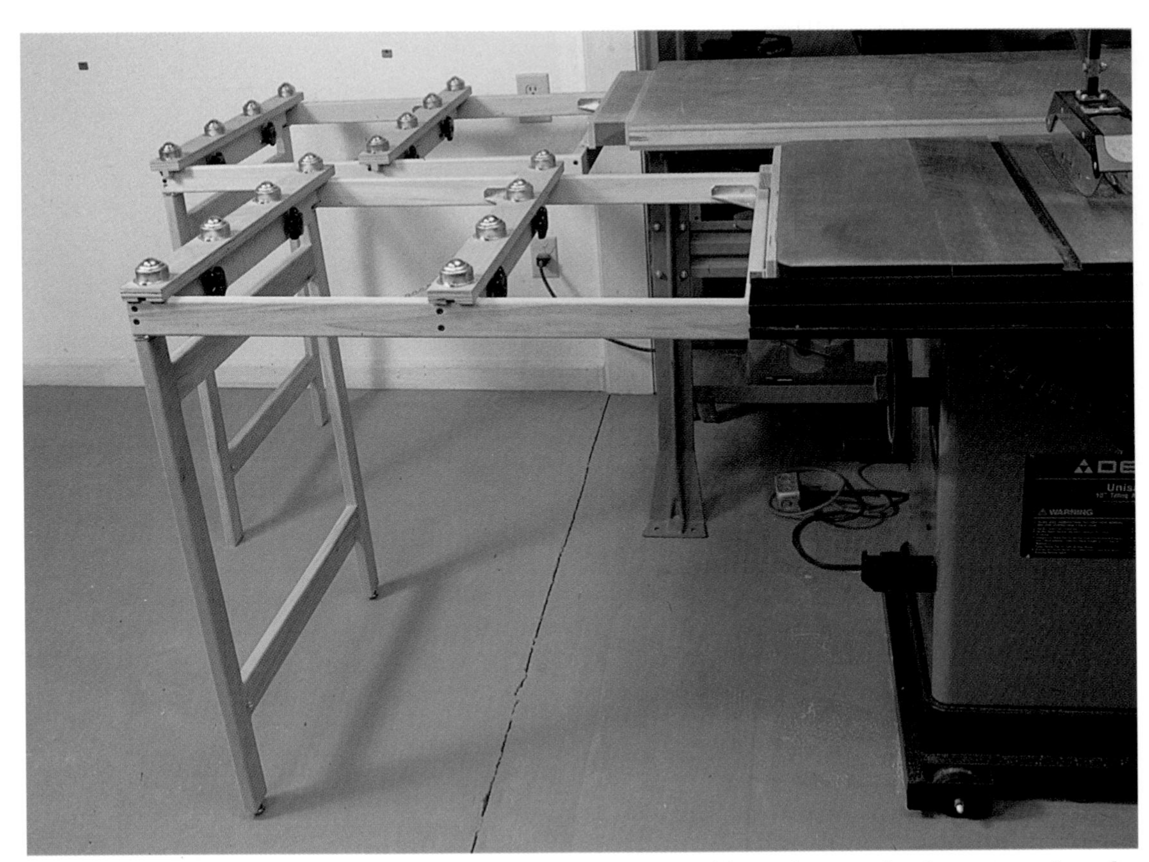

A roller-ball extension table makes cutting large panels safer and easier. *Unlike long tube rollers, roller balls won't pull stock out of line as it goes through the saw.*

Bob Gabor is an amateur woodworker in Pittsboro, N.C., and a member of the Triangle Woodworkers Association.

From *Fine Woodworking* (May 1995) 112:78-79

No-Hassle Panel Handling

Moving and storing sheet goods doesn't have to be backbreaking labor

by Skip Lauderbaugh

When you need big, flat panels that are stable, smooth and ready to be cut, you just can't beat sheet goods. But moving plywood, melamine or medium-density fiberboard (MDF) is a backbreaker. A single 4x8 sheet of ¾-in. MDF weighs almost 90 lbs., and it's terribly awkward to maneuver, especially by yourself.

I used to think that schlepping panels by hand was a necessary evil in my cabinetmaking business. Like many small-shop owners, I didn't have the space or the budget for material-handling equipment like a forklift. I stored panels near my saw in a stack. But it seemed whichever panel I wanted was always buried at the bottom of the pile. The day I needed a panel that was under 30 sheets of melamine, I just knew there had to be a better way.

It was time to stand back and analyze

Photos: Alec Waters

1. Platform is at a comfortable height for unloading. *The author slides plywood from his truck to the platform and tilts the panels up to the stack. He doesn't have to lift the full sheet.*

2. Bolsters let you leaf through sheet goods. *When sorting through panels, two bolsters act like buttresses to support sheets at the front of the stack. The bolsters adjust by sliding and locking in tracks in the top of the platform (below). An overhead rack holds small cutoffs.*

my entire panel-handling process—from unloading the truck to pushing panels through the saw. My goal was to devise a way for one person to unload, store, sort and move panels to the saw, using the least possible effort. So I came up with a storage system built around a low platform.

Panel-storage system saves labor, space and time

When I began studying how I had been moving sheet goods, I realized how inefficient I'd been. So I designed a panel-storage system to achieve five basic objectives:

- provide easy access to panels
- minimize lifting of entire sheets
- work at safe, comfortable positions
- organize panel cutoffs
- make the most of my floor space

At the heart of the panel-handling system is a 4-ft. by 10-ft. platform. The top of the

3. The right space between sheets and the saw—*After the author selects a panel, he pulls it end first from the stack. The platform is 6 ft. from his saw so that both ends of the sheet can be supported.*

4. Panel supported at start of cut—*The placement of the platform allows easy access to the saw and enables one person to move and cut panels. Leaving the front edge on the saw, the author feeds a panel into the blade by holding the unsupported back corner.*

From *Fine Woodworking* (September 1995) 114:82-84

***Stored sheets are easier to lift**—When the author has to carry a full sheet (top), he lifts it upright to keep his back straight. A cutout gives his hand clearance to grab the sheet's lower edge.*

***Drawers make use of floor space**—A two-part assembly table rolls under the platform when not in use. Aligned by biscuits and clamped together, the table has slide-out bins in back (bottom).*

platform is 24 in. above the floor, which is easy on the back for those rare times that I have to lift an entire sheet. The top is also at the right height for sliding sheets directly off the tailgate of my truck. And by standing on the platform, I can leaf through panels or reach up to my overhead cutoff rack. Connected to the top are two panel supports (I call them bolsters) that slide in tracks. The bolsters can be removed for loading panels or adjusted to fit the stack of sheets as it grows or shrinks.

I store panels with the long edges on the platform and the faces leaning against the wall. To sort through the stack, I lean unwanted panels against the bolsters and leaf through the rest like pages in a book. The end of the platform is 6 ft. from the front of my saw, providing plenty of cutting room. But I can still rest an end of a sheet either on the saw table or on the platform. To maximize floor space, I built two low assembly tables that roll under the platform.

The no-sweat panel shuffle

The beauty of the panel-handling system is that I almost never have to lift a full sheet. I either slide the panel or lift only one end. Photos 1 to 4 on pp. 10-11 show my typical panel-moving sequence. If I do have to lift a sheet off the platform, a cutout makes it easy (see the top photos).

There are only four elements in my panel-storage system: the platform, the bolsters, the cutoff racks and the assembly tables. I'll briefly explain how I built the platform, but I'll leave the specific measurements and details up to you. If you don't have headroom for overhead racks, for example, you can mount them somewhere else.

***The platform**—*The platform must be sturdy and big enough to hold 4x8 sheets. I designed the framework so I'd get the most storage area from the dead space underneath. I used 4x4s and 2x4s for the frame and secured it to a 10-ft.-long ledger I bolted to the wall. I anchored each leg of the platform to the floor.

The top of the platform has a pair of grooves running across the width, which serve as tracks for the adjustable bolsters. I bored 7/8-in.-dia. holes in the grooves every 6 in. to register the bolsters (the bolsters have alignment pins on the bottom) at various preset positions. And I sleeved the holes with short pieces of Schedule 40 PVC pipe to keep the holes from wearing and to keep the pins clean.

I let in and epoxied 1/8-in.-thick steel bars along both edges of the grooves to create lips to secure the bolsters (see the far right photo on p. 11). The bars protrude 3/8 in. into the groove, leaving a 5/8-in. gap between the bars. The top of the platform—3/4-in. plywood covered by 1/4-in. tempered hardboard—is screwed to the frame. On the wall behind the platform, I attached a 1/2-in. sheet of particleboard, so the panels have a flat surface to lean against.

***The adjustable bolsters**—*The bolsters measure 32 in. tall and are 9 in. wide at the bottom, tapering to 2 in. at the top. The cores are made of solid wood with 3/4-in. plywood gussets glued and screwed to the sides. At the bottom of each bolster are two pins. One pin is a 1/2-in. carriage bolt that fits into the track holes to align the bolster; the other pin, also a 1/2-in. carriage bolt, is inverted and has a 1-in. flat washer under the head. This pin prevents uplift on the toe of the bolster as sheets are loaded against it. The pins are height adjustable, so they engage both the holes and the lips of the track. Adjust the pins so they fit snugly in the tracks. Then carefully lean sheets against the bolsters to make sure they'll hold. You don't want a stack of sheets to crash against your legs later.

***The cutoff racks**—*Because I have a nice high ceiling, I made a rack above the platform for various sized cutoffs. The overhead rack is divided into three sections. The left section holds 12-in.-wide pieces, the center 18-in.-wide pieces and the right 24-in.-wide cutoffs. I located the bottom edge of the rack 62 in. above the platform to allow for 5-ft.-wide sheets and metric-sized plywood on the platform. The rack is attached to a ledger bolted to the wall. To the right of my platform is a storage rack that I use for wide cutoffs and long rippings. I can also use this area to store full panels vertically.

***The roll-out assembly tables**—*The space underneath the platform was the perfect place for storage-drawer units that also serve as cabinet-assembly tables. The two units are on wheels. They can be joined together to make one large surface, and when both tables are rolled under the platform, four drawers face out (see the bottom photos at left). I keep fasteners and hardware in these. When I pull the tables out, there are plastic crates in the back where I store power tools. The crates slide out on pull-out shelves. □

Skip Lauderbaugh is a sales representative for Blum hardware and a college woodworking instructor. His shop is in Costa Mesa, Calif.

Making a Sliding Saw Table

Smooth and precise crosscuts for less than a hundred bucks

by Guy Perez

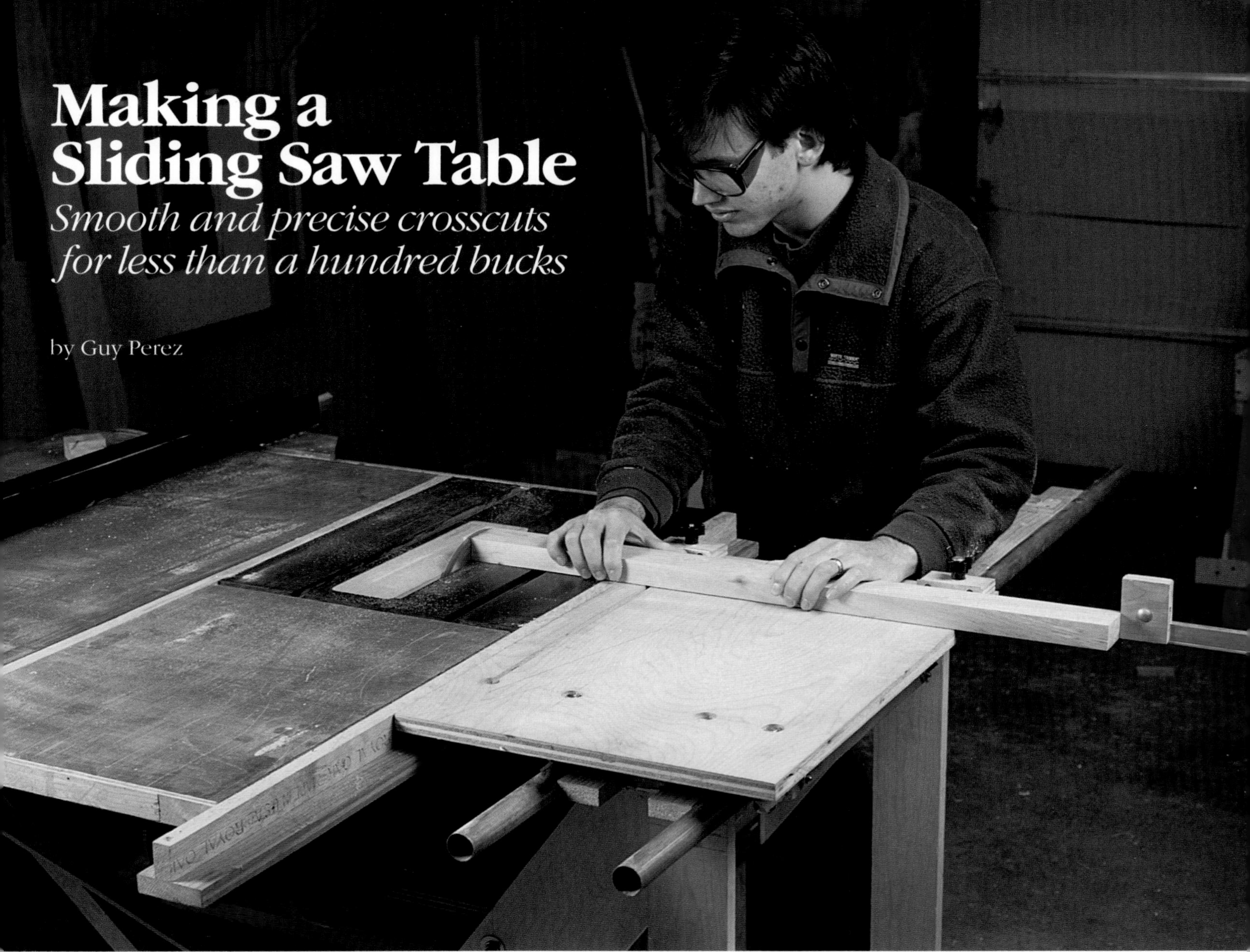

A sliding table improves crosscutting and mitering. *Guy Perez made this sliding table to extend the usefulness of his old Sears contractor's saw. Using lightweight everyday construction materials like plywood, pine and aluminum angle, Perez built the table with an adjustable fence, which makes the jig ideal for multiple crosscutting and for mitering.*

Until I came upon a 9-in., used tablesaw (a 1937 Craftsman model), I cut all my wood with an 8¼-in. circular saw aligned by a pair of shopmade guides. But even the Sears tablesaw still lacked a stand, table extensions and a miter gauge. So I set out to bring my bargain saw up to a higher standard.

The first additions I made to the saw included a stand, table extensions and a T-square fence, which allows me to rip stock up to 32 in. wide. These improvements served me well through several furniture projects, but I continued to crosscut with my circular saw and guide instead of using a miter gauge. As I saw it, standard miter gauges have three weaknesses: Their bars often fit loosely in the miter slots, they don't support long pieces well and they're ineffective for crosscutting wide pieces, especially sheet stock. I decided a sliding table would solve all of those problems.

But I found that most commercial sliding tables cost in excess of $350. The ones I looked at also failed to address another constraint I had—scarcity of shop space. So I built a scaled-down sliding table (see the photo above) that has a 32-in. crosscut capacity. The table cost me less than $100, but it performs comparably to the expensive commercial models. It was fairly easy to build, too, and I can still roll my entire tablesaw out of the way to save space.

How the table slides

Like a few of its store-bought cousins, my sliding table rolls on precision bearings that are guided by steel rails. The whole assembly (see the drawing on p. 14) consists of four main components: an 18-in. by 24-in. plywood table, an 18-in.-long carriage that has four pairs of bearings, two 5-ft.-long tubular guide rails and a 60-in.-long aluminum crosscutting fence. Similar to one of Robland's sliding tables, the rails are spaced about 6 in. apart and are mounted left of the saw table. Drawing detail A on p. 14 shows the wooden frame I built to support the rails. You can easily modify the frame to suit the saw you have, as long as the sliding table is level with and travels parallel to the saw table.

Constructing the table

Before you start to build a sliding table, there are a couple of things worth noting about aluminum. First, when buying aluminum bar or angle, check out recycling centers and salvage yards because they usually don't require a minimum quantity or charge the premium that metal-supply shops often do. Second, you can cut aluminum to length on a tablesaw fitted with a carbide-tipped blade. But be sure you don't let the hot chips touch your skin.

Photos except where noted: Alec Waters

Sliding table anatomy

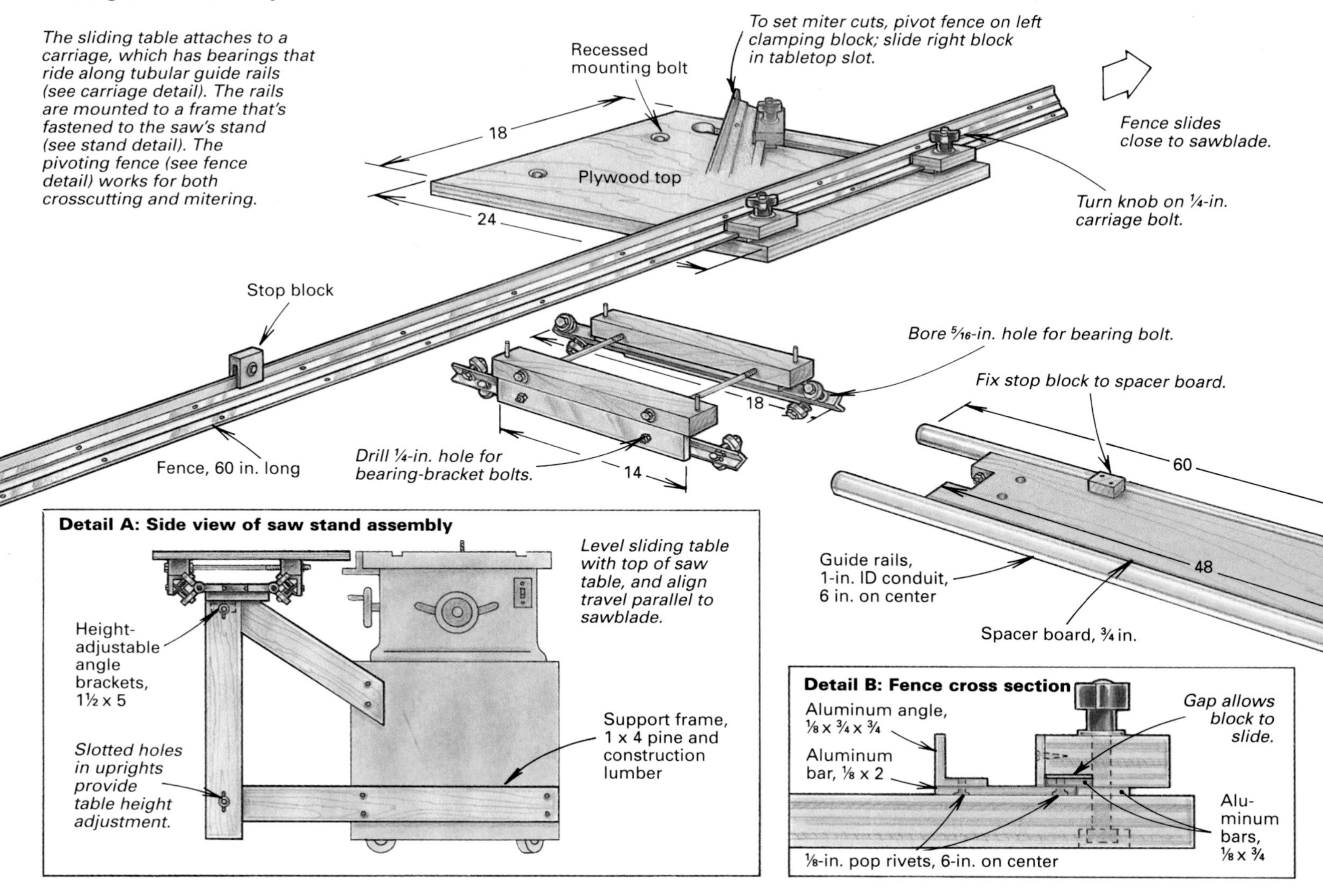

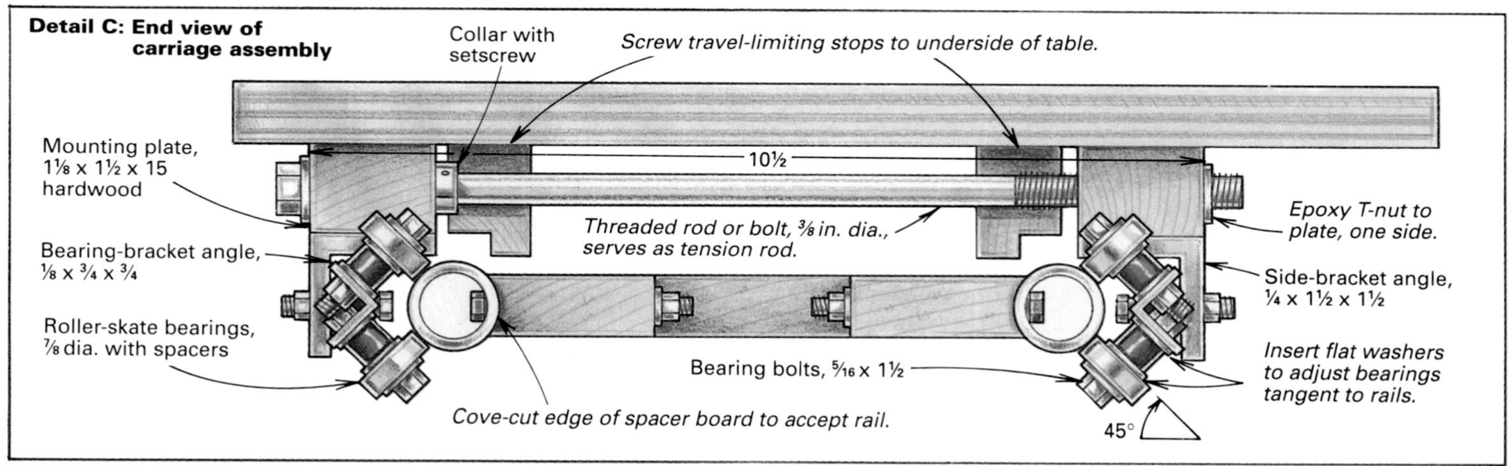

Carriage—The carriage is the heart of the sliding table, so I built it first. It acts much like a sliding dovetail joint, in which a pin is held by the tapered dovetail groove, thus restricting side-to-side or up-and-down movement. In my sliding table, the pin is replaced by guide rails captured by opposing pairs of bearings (see drawing detail C). By mounting the bearings 45° above and below the plane of the rails, I restricted both lateral and vertical carriage motion while allowing the table to roll forward and back.

I mounted two pairs of bearings and spacers to each of two bearing brackets (¾-in.- by 18-in.-long aluminum angles), putting one bearing and spacer pair at each end (see the photo at right on the facing page). The bearings and spacers I used are intended for skateboard wheels and are available at most sporting-goods stores and hobby shops for around $16 for eight bearings with spacers. I secured each bearing bracket to the vertical leg of the side brackets (1½-in.- by 14-in.-long aluminum angles) so that each leg of the bearing-bracket angle is 45° off the horizontal plane of the rails. I found it easiest to first drill the mounting holes in the 1½-in. side-bracket angles. Then, with the ¾-in. bearing-bracket angle clamped in place, I drilled its corresponding holes.

The side-to-bearing bracket assemblies are held together by two tension rods (I used hex bolts, but ⅜-in. threaded rod will also do).

Drawing: David Dann

Photo: author

***The guide system is strong and accurate** (above). Adapting a design from commercial-grade sliding tables, Perez made a roller-skate-bearing carriage and attached it to the underside of the table. The carriage is guided by a pair of tubular rails.*

***Crosscutting fence, extension and outfeed tables** (left) add to the accuracy, safety and versatility of the author's saw. Reduced friction and vibration of the sliding table are real assets when crosscutting long stock and sheet goods or mitering pieces—operations made possible by the added accessories.*

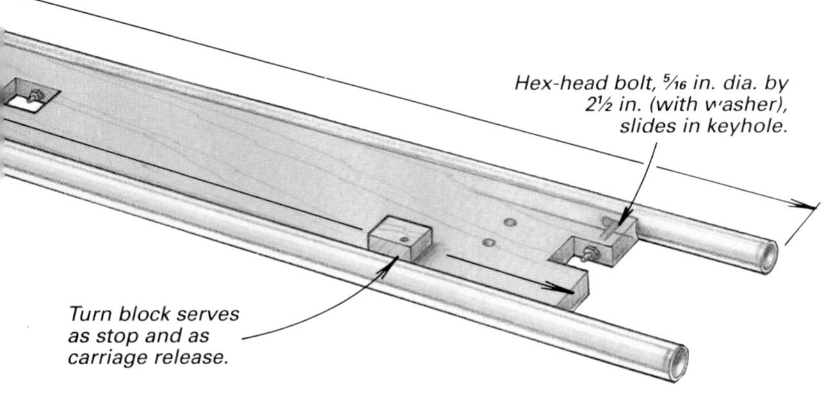

The tension rods allow the carriage to be precisely fit to the guide rails. The rods are fastened so that they turn freely in a mounting plate and are fixed to another plate with a T-nut that's epoxied in place. The 1⅛-in.-thick hardwood mounting plates are screwed to the top of each 1½-in. angle. Four T-nuts with bolts near the ends of the mounting plates secure the table to the carriage.

Guide-rail assembly—I built the guide-rail system to fit between the four pairs of carriage bearings. For the rails, I used 1-in. ID electrical conduit, which goes for about $4 per 10 ft. at building supply stores. I bolted a 6-in.-wide strip of ¾-in. particleboard as a spacer between the rails to stiffen them and keep them parallel. An easy way to determine the exact width of the spacer board is to set the tension rods, so there is ½ in. of adjustment either way. Then hold the rails between the bearings, and measure the distance between the rails. Because the spacer board is cove cut on both edges to accept the tubing, add ¼ in. to the measurement. In each of the rails, I cut three keyhole-shaped slots by drilling sets of holes—each set with a ½-in. hole overlapping a ¼-in. hole. The slots let me insert the hex bolts that fix the rails to the spacer board. I screwed a block at the rear of the spacer board to limit the table's travel. Then I added a turn block at the front (see the photo at left above), which lets me release the carriage from the guide rails.

Support frame—To make the frame that supports the guide-rail system, I used eight board feet of 1x4 pine. I made the frame so I could easily get to all my saw's controls. Two 1½-in.- by 5-in.-long aluminum-angle brackets hold the rail assembly to the frame's uprights. Slotted holes in the uprights and oversized holes in the brackets provide the means for height adjustment.

Final adjustments

The carriage should be fit to the guide rails before you mount the sliding top. After adjusting the tension bolts so the bearings fit the rails, you should check two other things to ensure that the table will slide properly. First, the bearings should contact each guide rail tangentially. Second, each mounting plate should be parallel to the rail below it. I adjusted my bearings by using flat washers to get the proper spacing. To ensure parallel travel (vertically with respect to the rails), I used a combination square and set each mounting plate the same distance above the rails at the front and rear. Initially my carriage was misaligned. To fix this, I elongated the bearing-to-side-bracket holes using a rat-tail file, and then I repositioned the brackets.

I made the final carriage adjustments with the sliding table in place. To align the sliding table with the top of the saw table, I clamped 4-ft.-long straightedges across the front and rear of the saw table. I leveled the sliding table up to the straightedges and tightened the frame's height-adjustment bolts. Next I clamped a board to the sliding table, perpendicular to the rails. I drove a finishing nail in one end of the board, leaving it about ¼ in. proud. As I moved the table to and fro (with the saw unplugged), I measured from the nail head to the blade, both front and back. Once I was sure the table was parallel, I snugged up all the mounting bolts. Then I screwed travel-limiting stops to the underside of the table in line with the spacer-board blocks. To position the stop blocks, I rolled the carriage and marked limits for the table's normal movement.

Finally, I equipped my sliding table with a 60-in. crosscut fence (see drawing detail B on the facing page). Because the fence is adjustable, I can set it for mitering, and I can position it to support a workpiece right up to the blade. Fitted with an adjustable stop, my fence and sliding table made quick work of cutting slats for a crib and couch I was building. □

Guy Perez is completing his dissertation in political philosophy at the University of Wisconsin in Madison. In his spare time, he builds woodworking machines to improve his furnituremaking.

Roller-topped drawers increase outfeed table capacity. *By extending the bottoms of two drawers at the back of his tablesaw, Frank Vucolo created a place to mount outfeed rollers. Here, he opens one drawer to rip a piece of 6/4 mahogany.*

Drawer slide alignment is important. *With the outfeed table flipped, the author positions a slide before he screws it to the poplar rail. Precise alignment ensures smooth operation of the outfeed rollers. A leg socket is below the square.*

Shopmade Outfeed Table

Extend your tablesaw's reach for sheet stock and ripping

by Frank A. Vucolo

In my small shop, ideal concepts are often compromised by the reality of limited space. My design for an outfeed table is a classic case in point. I started out thinking big. Ideally, I wanted the outfeed surface to extend 48 in. from the back of my tablesaw, so I would no longer have to set up and then reposition unstable roller stands. My ideal was quickly squashed, however, when I realized I couldn't dedicate that much permanent floor space. I need the space behind the saw to store my planer and router table when I'm not using them.

After some careful measuring, taking into consideration where I would locate all the machines, I concluded that the outfeed table should extend 30 in. from the back of the saw. But I still needed more support to rip long stock and to cut sheet goods.

While I was pondering possible solutions, I started to think about rollers that could extend off the back of the fixed table and then retract into it when they weren't needed. Then I remembered how amazed I was at the strength of Accuride's extension drawer slides (150-lb. capacity) when I had used them for file drawers in a desk pedestal. After a little more head scratching, nudged along by a couple of cups of coffee, I decided to incorporate the slides into a pair of drawers with rollers mounted on the front of them for the outfeed table (see the photo at right above). Now I

Photos: Alec Waters

Outfeed table assembly

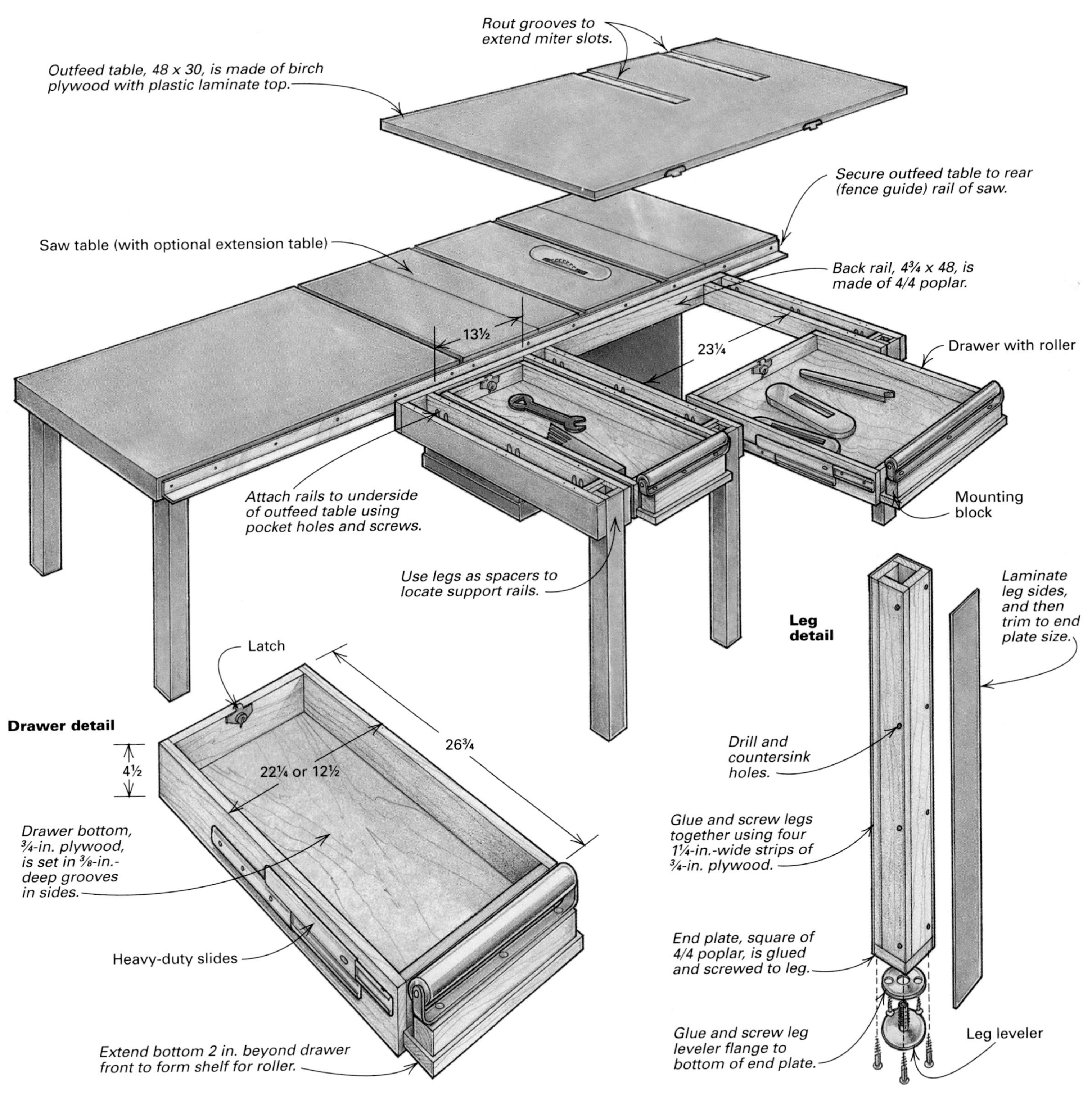

simply open a drawer to get an additional 24 in. of outfeed surface when I'm ripping long boards or cutting sheet stock.

Design and materials

Allowing an extra 1 in. for the extension rollers and the drawer slide action, the outfeed table is designed to support work up to 55 in. from the back of the saw table. With the drawers in the closed position, only 30 in. of floor space behind the tablesaw is committed. I made the drawers different widths so that I have various outfeed options, and I extended the drawer bottoms out in front of the drawers. This way, I have a place to mount the rollers (see the detail above). As a bonus, I get two drawers for storing saw accessories. And because the rollers are an integral part of the outfeed table, they are adjusted precisely in relation to the tabletop.

I constructed the outfeed table's top, legs and drawer bottoms out of ¾-in. birch plywood. The under-table support rails are made from 4/4 poplar, as are the drawer sides, fronts and backs. For added protection and to give a nice slick surface, I covered the legs and top with plastic laminate.

To complete the material requirements, I bought the following hardware: two metal rollers, one 13 in. long and one 22 in. long (Wilke Machinery Co., 3230 Susquehanna Trail, York, Pa. 17402;

Drawing: David Dann

800-235-2100), two sets of heavy-duty drawer slides (I picked up Accuride's file-cabinet model from The Woodworkers' Store, 21801 Industrial Blvd., Rogers, Minn. 55374; 800-279-4441), three leg levelers (available from Woodworker's Supply Inc., 1108 North Glenn Road, Casper, Wyo. 82601; 800-645-9292) and a couple of latches (window sash locks), which I bought at a local hardware store. When you're determining the size of your drawers, keep in mind that the slides come in 2-in. increments, 12 to 28 in. long.

Making and mounting the table

To build the outfeed table, first determine the overall size (mine is 48x30), and then cut the tabletop out of plywood. Temporarily mount the plywood to your saw, and level it using braces. This is so you can determine the length of the three legs. Measure each leg separately, and allow some room (½ in. or so) for height adjustment. The leg levelers will take up the play. Disassemble the table, and then fabricate the legs, as shown in the drawing detail on p. 17, including the plastic laminate.

With all three legs complete, lay out the support rail locations on the underside of the plywood top. Approximate the two different widths of the drawers plus their slides. Rip and crosscut the poplar pieces to size, and begin fixing the members to the plywood. I drilled pocket holes and then glued and screwed the rails in place. Start at one end, then use an assembled leg as a spacer to set the second rail. Next do the other end of the table, using another leg as a spacer. Set the two center rails in a similar fashion. Then attach the rear rail across the ends of the support rails. Also, cut and attach blocks behind each leg using the leg as a guide.

Mount the carcase portion of each drawer slide to the rails (see the photo at left on p. 16). Make sure you position all the slides the same distance from the bottom of the table. I used the rails as a reference. The drawers must be perfectly parallel to the top. While you have the table flipped, laminate the sides of the top, and trim them with a flush-trimming bit in a router. Turn the table over, so you can laminate and flush-trim the top.

Now temporarily mount the legs, and align the laminated table to your saw exactly as it will be positioned in use. Carefully mark the position of the miter slots on the top. Determine the depth of the grooves by referencing off the tablesaw. If you have a T-slot or dovetail-shaped miter-gauge runner, lay out the slots so that they will be a bit wider than the widest (bottom) part of the tablesaw slot. The outfeed table slots will be for clearance only.

Remove the outfeed table. Run the miter gauge all the way past the blade, so you can find the length of the runner as it hangs off the back of the saw table. Mark this length plus a bit extra onto the outfeed tabletop. If you use sliding jigs, like a crosscut box, check that their runners will work in the laid-out slot, too.

Using a straight bit and your router, cut the grooves in the surface of the outfeed table. A straightedge can be used to guide the router. But don't try to cut the whole depth in one pass. It's better to make two or three passes, removing a little at a time. Soften all the corners of the laminated top using a fine file. Also, ease the edges of the miter-gauge slots, and feather the edge that will go against the tablesaw. This will ensure that workpieces won't get hung up as they slide from the tablesaw onto the outfeed table.

How you mount the outfeed table to the saw will depend on the type of saw and fence guide rail you have. You can use angle brackets or drill directly into the rail. After you have the outfeed table in its approximate position, use a straightedge and a level to adjust the screw feet until the outfeed table is lined up to the saw table (see the photo at left).

***Level the outfeed table to match the saw table**—After Vucolo secured the outfeed table to the rear guide rail of his saw, he turns the leg levelers (screw feet) to line up the two surfaces.*

***Pocket holes and screws join drawer boxes**—After temporarily clamping a drawer back, the author drives three screws into the sides using a flexible-shaft extension for his drill.*

Adding the drawers and extension rollers

The drawers should have a ¾-in. plywood drawer bottom extending 2 in. beyond the front of the drawer. This will provide enough rigidity for the extension rollers (see the drawing detail on p. 17). To receive the bottom, I plowed a ⅜-in.-deep groove down the inside of each drawer side using a dado blade in my tablesaw. After I glued and screwed the bottom to each drawer, I butt-joined the front and back pieces together using pocket holes and screws (see the far left photo). Then I attached the other part of the drawer slides to the outsides of the drawers.

It's critical that the rollers are mounted at the correct height. They should be at, or just barely above, the outfeed surface; they need to roll freely, without disrupting the travel of a workpiece. To get the proper height, I mounted the rollers using spacer blocks. First I set the roller on the shelf created by the extended drawer bottom. Then I measured from the top of the roller to the tabletop. I cut the block a bit oversized and then planed it down to thickness. If the roller is not parallel to the outfeed top and you can't adjust the drawer slides enough, taper the blocks slightly with the plane until the top of the rollers are level with the table. Finally, install a latch on the inside back of each drawer, so you can lock them in the open position. □

Frank Vucolo builds furniture for his home in East Amwell, N.J.

From *Fine Woodworking* (September 1994) 108:74-76

Router Fixture Takes on Angled Tenons

Versatile device ensures tight joints every time

by Edward Koizumi

We live in a turn-of-the-century Arts-and-Crafts house, so it seemed quite natural to furnish it with pieces from that era. My wife bought a pair of Mission armchairs a couple of years ago to go with a 9-ft.-long cherry table I'd built for our dining room. Six months later, she bought two side chairs. It would be a while before we could afford a full set. Within earshot of my wife, I heard myself say, "How hard could it be to make these?"

"Oh, could you?" she asked.

"Sure," I said. The chairs looked straightforward enough, just a cube with a back. Upon closer examination, I realized that the seat was slightly higher and wider in the front than in the back. For the first time, I was faced with compound-angled joinery. I thought about dowels, biscuits and loose tenons, so I could keep the joinery simple, but I wasn't confident in the strength or longevity of these methods.

I wanted good, old-fashioned, dependable mortise-and-tenon joints. After some thought, I decided an adjustable router fixture would be the simplest solution that would let me make tenons of

Bottom photo: Boyd Hagen

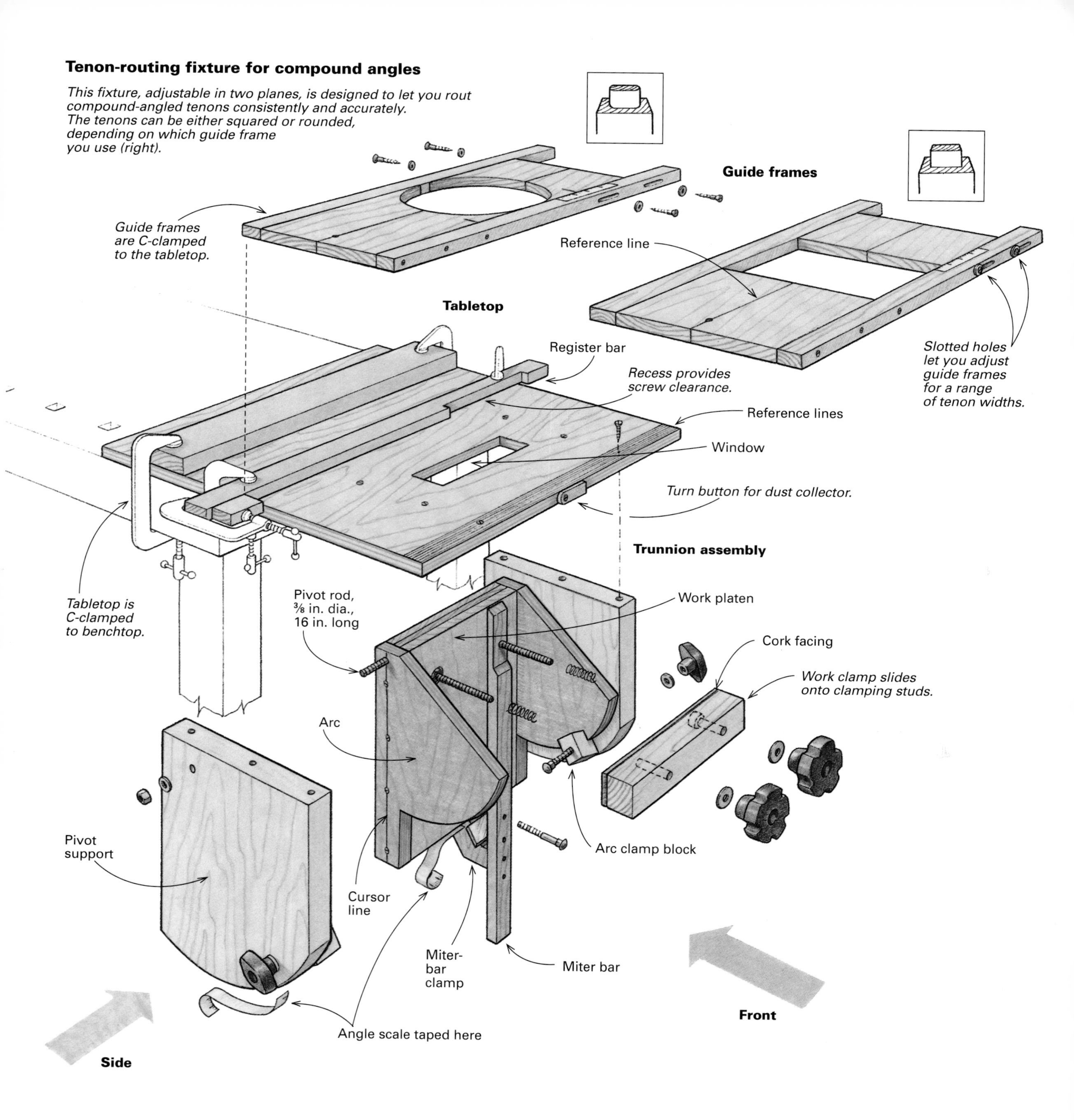

widely varying sizes and angles (see the photos on p. 19).

The fixture I came up with is as easy to set up as a tablesaw. In fact, there are some similarities (see the drawing above). The workpiece is held below a tabletop in a trunnion-type assembly that adjusts the tilt angle (see the bottom photo on p. 19). For compound angles, a miter bar rotates the workpiece in the other plane. The fixture can handle stock up to 2 in. thick and 5 in. wide (at 0°-0°) and angles up to 25° in one plane and 20° in the other. This is sufficient for chairs, which seldom have angles more than 5°.

To guide the router during the cut, I clamp a guide frame to the fixture over the window in the tabletop (more on positioning it later). And I plunge rout around the tenon on the end of the workpiece. The guide frame determines the tenon's width and length, as well as whether the ends will be square or round (see the photo on p. 23). I made two frames, both adjustable, one for round-cornered tenons, the other for square tenons.

The fixture and guide frames took me just over a day to make, once I'd figured out the design. Then I spent about an hour align-

Drawings: Heather Lambert

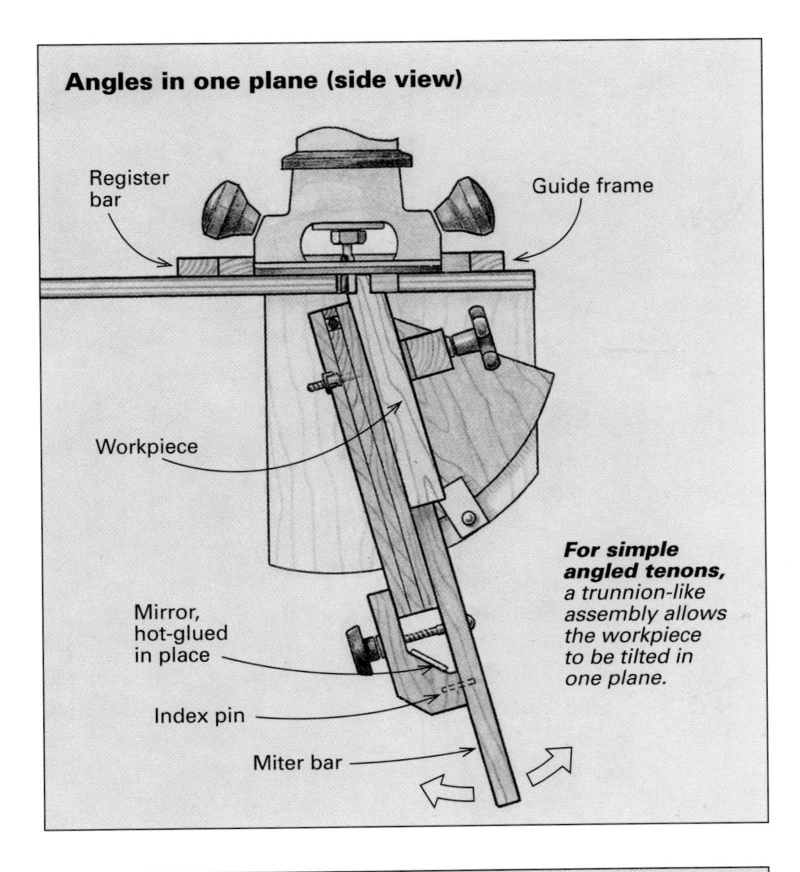

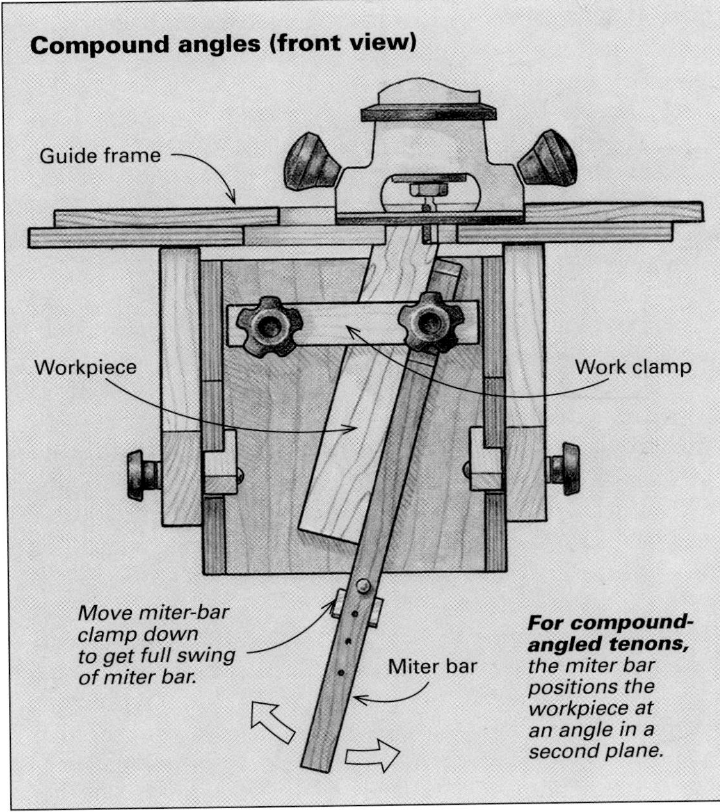

ing the fixture and making test tenons in preparation for routing the tenons on the chair parts. The fixture worked just as planned and allowed this relatively inexperienced woodworker to produce eight chairs that match the originals perfectly.

Making the fixture and guide frames

The fixture is simple to build. It consists of only two main parts, the trunnion assembly and the tabletop. The trunnion assembly (see the drawing on the facing page) is essentially a pair of arcs nestled between two pivot supports. Between the two arcs is a work platen, or surface, against which I clamp the component to be tenoned. There are other parts, but basically, the fixture is just a table to slide the router on and a movable platen to mount the workpiece on.

I built the fixture from the inside out, beginning with the work platen (see the drawing on the facing page). Because I didn't have any means of boring a 10-in.-long hole for the threaded rod on which the arcs pivot, I dadoed a slot in the platen and then glued in a filler strip. Next I located, center punched and drilled the holes for the T-nuts and retaining nuts that hold the clamping studs in place. Center punching ensures that the holes are exactly where they're supposed to be, which is important for a fixture that's going to be used over and over again. I center-punched the location for every hole in this fixture before drilling.

Before attaching the clamping studs to the work platen, I made the arcs, which go on the sides of the work platen. I laid out the arcs (and the pivot supports) with a compass, bandsawed and sanded the arcs, and drilled a hole for the pivot rod through the pair. I glued and screwed the arcs to the platen. After giving the glue an hour or so to set, I tapped the T-nuts into the back of the work platen, screwed in the clamping studs and twisted on retaining nuts, which I tightened with a socket and a pair of pliers.

I made the pivot supports next. Then I cut a piece of threaded rod 16 in. long and deburred its ends with a mill file. I slipped the threaded rod through the pivot supports, arcs and work platen, capped it at both ends with a nut and washer, and made and attached the arc clamps (see the top drawing at left).

Then came the tabletop. I cut it to size, cut a window in it and marked reference lines every ⅛ in. along the front edge for the first 2 in. With the tabletop upside down on a pair of sawhorses, I put the trunnion assembly upside down on the underside of the tabletop. Then I positioned the front of the pivot supports against the front edge of the tabletop and made sure the work platen was precisely parallel to the front edge and centered left to right. That done, I drilled and countersunk holes for connecting screws through the tabletop into the pivot supports. I glued and screwed the pivot supports to the tabletop.

Then it was time to make the miter bar, miter-bar clamp and the work clamp (see the drawings at left). The mirror on the miter-bar clamp makes it easy to read the angle scale from above. I faced the work clamp with cork to prevent marring workpieces and counterbored it to take up the release springs. The release springs are a nice touch. They exert a slight outward pressure on the work clamp, causing it to move away from the platen when loosening the knobs to remove a workpiece.

The guide frames—Now for the guide frames, which clamp to the tabletop and limit the travel of the router. I made the frames adjustable lengthwise to handle a variety of tenoning situations. But their width is fixed. To determine the width of the frames, I added together the desired tenon width, the diameter of the bit I was using and the diameter of the router base. If your plunge router doesn't have a round base, you should either make one from acrylic or polycarbonate (you can cut it with a circle-cutting jig on a bandsaw), or buy an aftermarket version. I screwed the frame together in case I need to alter the opening later (for a new router bit, for example). I marked a centerline along the length of the frame on both ends.

Initial alignment

Before I could use the fixture, I had to get everything in proper alignment and put some angle scales on it. I printed out some an-

From *Fine Woodworking* (September 1995) 114:77-81

SETTING UP FOR ANGLED TENONS

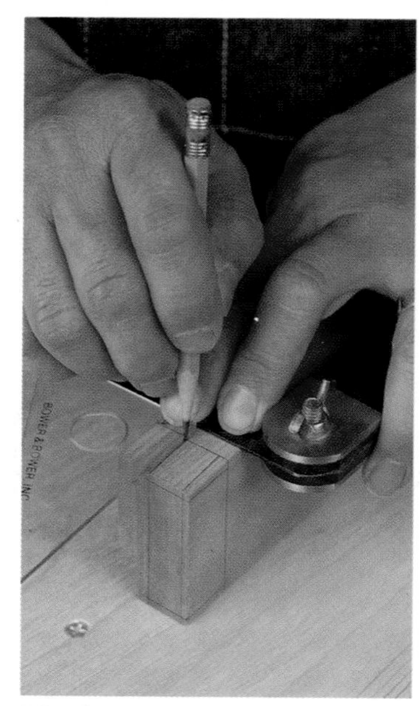

Mark out the tenon on a test piece. *The test piece should be the same thickness and width as the actual components, but length isn't important.*

Make the workpiece flush with the tabletop. *The author uses a piece of milled steel, but the edge of a 6-in. ruler would work as well.*

Make a pattern. *An outline of the tenon traced on acetate helps align the guide frame for cutting any tenons of the same size.*

gle scales from my personal computer and taped them to my fixture with double-faced tape. But a protractor and bevel gauge also will work just fine to create angle scales for both the tilt angle and the miter angle.

To align the parts of the fixture, I flipped it upside down on the end of my bench and clamped it there. I used a framing square to set both the work platen and the miter bar at 90°, sticking the blade of the square up through the window of the tabletop and resting the tongue of the square flush against the inverted face of the tabletop. Then I stuck the angle scales on the two pivot supports and on the bottom of the work platen.

Routing test tenons

Next I routed test tenons with the fixture set at 0°-0°. I positioned the guide frame parallel to the front edge and centered on the window in the tabletop and clamped it to the fixture. I clamped a test piece the same thickness and width as the actual component in the fixture, with one end flush with the top surface of the tabletop. To do this, I brought the test piece up so that it just touched a flat bar lying across the window (see the near left photo). I set my plunge router for the correct depth and routed the tenon clockwise to prevent tearout.

I made a test mortise using the same bit I planned to use for the mortises in the chair. The fit wasn't quite right. So I adjusted and shimmed the frame until the tenon fit perfectly. If you rout away too much material and end up with a sloppy tenon on your test piece, you can just lop off the end and start over.

Once I had a tenon that was dead-on, I made an acetate pattern that allowed me to position the guide frame accurately for all tenons of the same size, regardless of the angle. I cut a heavy sheet of acetate (available at most art-supply stores) so that it would just fit into the guide-frame opening. I marked a centerline along the length of the acetate that lines up with the centerline down both ends of the guide frames. I also indicated which end was up and where the acetate registered against the guide frame. Then I put the test piece with the perfectly fitted tenon back into the fixture, laid the acetate into the opening in the guide frame and traced around the perimeter of the tenon end using a fine-tip permanent marker.

Routing angled tenons

With the pattern, routing angled tenons is pretty straightforward. I crosscut the ends of all the pieces I was tenoning at the appropriate angles and marked out the first tenon of each type on two adjacent sides, taking the angles off a set of full-scale plans. Then I extended the lines up and across the end of the workpiece (see the top left photo).

Having set the fixture to the correct angles, I brought the workpiece flush with the tabletop using a flat piece of steel as a reference (see the top right photo). Then I clamped the workpiece in place. Finally, I set the acetate pattern in the guide-frame opening and positioned the guide frame so that the pattern and the marked tenon were perfectly aligned (see the photo at left). With the guide frame clamped in place, I removed the acetate and routed that tenon. All other identical tenons needed only to be flushed up and routed. After the first, it was quick work.

There are pitfalls though. I found it important to chalk orientation marks on each workpiece. It can get confusing with two angles, each with two possible directions. And I had to be especially careful when routing the second end of a component. Make sure it's oriented correctly relative to the first. I messed up a couple of times and have learned to plan for mistakes by milling extra parts and test pieces. You might even end up with an extra chair.

Photos except where noted: Vincent Laurence

***Guide frame determines thickness and width of tenons.** The author keeps the router's base against the inner edges of the guide frame and routs clockwise to prevent tearout. Guide frames can produce round-cornered or square-cornered tenons.*

To get flat surfaces on curved parts so I could clamp them in the fixture, I saved the complementary offcuts and taped them to the piece I was tenoning. Or I could have tenoned first and bandsawed the curves later.

For pieces with shoulders wider than the bit I'm using to remove waste, I clamp a straight piece of wood—a register bar—against the guide frame (a small pocket for screw clearance may need to be made), as shown in the drawing on p. 20. That way I can rout most of the tenon, unclamp the guide frame, slide it forward (using the reference lines at the forward end of the tabletop to keep it parallel), clamp it down and then rout the remainder. I start the next piece in the same place and return the guide frame to the original position to finish the tenon. □

Edward Koizumi is a professional model maker in Oak Park, Ill.

***Set correctly, the fixture will yield tight joints,** whether the tenons are straight, angled or compound-angled. Here, the author tests the fit of a seat-rail tenon into a leg mortise.*

Stow-Away Router Table

Cantilevered frame clamps to bench quickly, stores in seconds

by Jim Wright

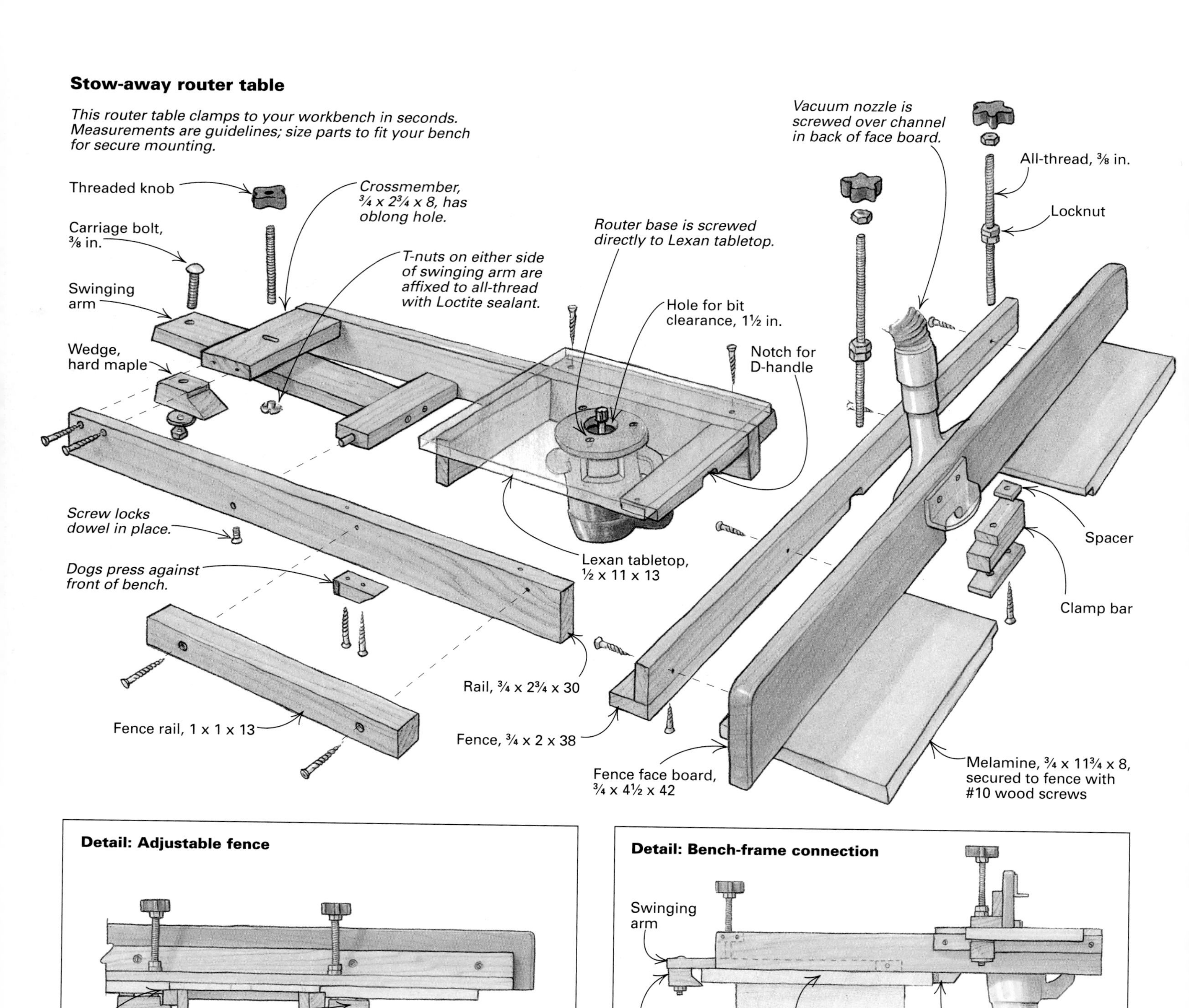

Drawing: Bob La Pointe

I'm afraid my early attempts at making a router table were nothing to write home about. The first three designs—a table with legs, a cabinet base and a table attached to my tablesaw—all ended up on the scrap heap. They were just too bulky, a fatal flaw when it came to my small shop and lack of storage space.

It finally occurred to me all I really needed was a simple router table that could be clamped to my workbench. Here's how it works: I attach my router to a plastic insert and drop it into the frame of the table; the router hangs over the front edge of my bench. A sliding fence rides on top of the frame and adjusts easily. The mass of the bench kept the router table from vibrating, but best of all, the whole assembly is compact and easy to store (see the drawing).

The key was in finding a way to clamp the assembly to my bench, so it wouldn't move. I did that by sizing the frame of the router table, so it spanned my bench exactly. Then I held the frame in place with a simple clamp made of a wooden wedge and a length of all-thread.

Building the frame for your bench and router

The first step is to decide how large a tabletop you need for your miter. Allow enough room for the knobs and handles on your router, and give yourself room to adjust the router when it's attached to the table. Just how much is enough depends on your router. I have a D-handle on my router, so I had to cut a relief in the framing to accommodate it. I made the frame for my router table of hard maple. The corners are fastened with #10 wood screws, so assembly is easy.

The router table stays in place because it grabs both the front and the back edge of my workbench. Attached to the rear of the router-table frame is a maple wedge that hooks over the back edge of my bench. Screwed into the bottom of the frame near the front of the table are two dogs that press against the front edge of the bench. I added 80-grit sandpaper on the inside faces of these pieces to give them a better bite. When I tighten the knob at the back of the frame, the wedge pulls up against the bottom of the bench and locks the frame into place. Because the dogs on the front of the frame are tight against the front of the bench, my router table really can't go anywhere.

The assembly should hold securely even with the clamp knob a little loose—that's important. If the fit between the router table and your bench is sloppy and the clamping mechanism were to fail, the table would fall on the floor. Not a nice picture: router, work and fingers all mixed together and heading for the deck.

Making the tabletop

The thick table is a piece of ½-in. Lexan from a dealer's scrap pile. The 11-in. by 13-in. piece cost me $10, but expect to pay more if you have a piece cut to size from stock. Lexan, a polycarbonate, cuts easily with a tablesaw, and trimming it to size is no problem. Other plastics may shatter or melt, but Lexan is lovely to work with. Phenolics also work well and are more rigid than Lexan. (For more on plastics in the woodshop, see *FWW* 105 p. 58.)

Big router-table performance in a benchtop package*—This shop-built router table sets up in seconds and stores easily when not in use. Securely clamped to your benchtop, it can do most anything more conventional router tables can.*

The base of the router dictates the layout for the mounting holes. Use a drill press with a spade bit turning at low speed (clamp the work) and light pressure to cut a 1½-in. hole in the center of the base (a piece of scrapwood beneath the work when drilling will keep the bit from jumping when it breaks through). The mounting holes are made with a twist bit, then countersunk. The Lexan is attached to the frame with four #10 wood screws.

Fabricating the fence

The fence is made to slide on the frame. Two clamps lock it in place. Attached to the right and the left sides of the fence are two pieces of melamine, which support the work as it's fed past the bit. The fence slides on or off in seconds. Once the router is mounted to the table, the table itself can be mounted or removed from the bench in 15 seconds—without changing the position of the fence.

The fence has a face board screwed to it with a channel in the back to create a duct for dust collection, as shown in the drawing. I added a plastic finger guard for safety. As a bonus, I find that the guard helps control the dust. □

Jim Wright is an amateur woodworker in Berkeley, Mass.

From *Fine Woodworking* (March 1995) 111:56-57
Photo: Vincent Laurence

Wall-mounted panel router is ideal for making quick dadoes. *Knowing his panel router had to save space, Skip Lauderbaugh mounted it to a wall at a comfortable height and angle. To build the jig, he used a router he owned and commercial hardware costing less than $100.*

Compact Tool Makes Dadoes a Snap

This panel router folds flat against a wall and is inexpensive to build

by Skip Lauderbaugh

Many of my cabinetmaking projects require panels that have dadoes, rabbets and grooves to allow strong, easy assembly. I've tried lots of ways of cutting these joints and have found that a panel router is the quickest and most accurate tool to use. Unfortunately, the expense of one of the commercial machines (up to $3,500) and the floor space it requires (up to 25 sq. ft.) is more than I can justify. As is often the case, however, once you have tasted using the proper tool for a particular job, using anything else becomes a frustrating compromise.

I had seen other shopmade panel routers (for one example, see Steven Grever's article in *FWW* #88, p. 48), but they lacked features I wanted and seemed complicated. So I set out to design and build my own version of a panel router. By simplifying the guide system and by using common materials and hardware (see the drawing on p. 29), I built a panel router for less than $100 (not including the router, which I already owned). And although this jig easily handles big pieces of plywood and melamine, the jig folds compactly against the wall when it is not in use.

Designing the panel router

Because the guide rails used in industrial panel routers often get in the way, the rails were the first things I eliminated on my design. The next thing was to orient the machine so that gravity would help feed the router into the work. Big panel routers are oriented horizontally, and they have the capacity to handle 36-in.-wide pieces of plywood. But because shelf dadoes in cab-

Photos except where noted: Alec Waters

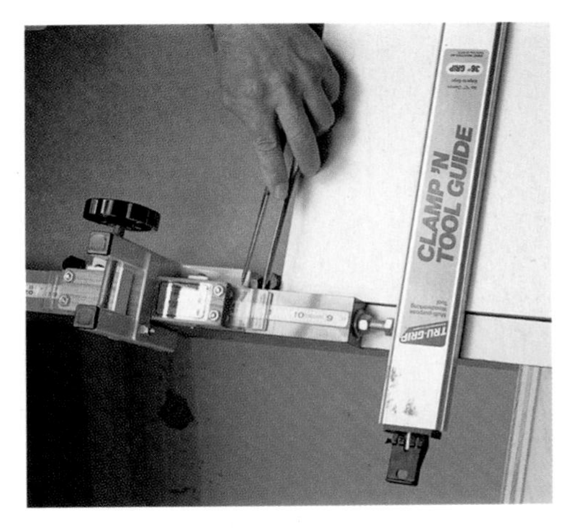

The fence's adjustable stop ensures perfect alignment. *A Biesemeyer micro-adjustable stop and measuring system precisely positions the left side of the work for each dado or groove. Lauderbaugh uses a pair of dividers to point out two cursors that indicate left and right limits of a cut.*

Channels align subbase and evacuate dust*—The underside of the router subbase reveals an inverted aluminum guide channel and a medium-density fiberboard bottom with dust-evacuation slots cut across it for the bit.*

The key to the router guide is interlocking aluminum track. *When the author discovered the edges of Clamp 'N Tool Guides nest and slide easily, he made them into a two-piece guide system: An inverted 21-in. piece is fixed to the router subbase, and another piece is clamped to the work.*

inets and cases are usually less than 3 ft. wide, I scaled things down a bit, and I situated the whole setup vertically. This orientation also saved considerable shop space. Then I came up with a clamp-on router guidance system, so I don't have to do any measuring or marking on a panel. Finally, I devised a router subbase that eliminates depth-of-cut adjustments when changing material thicknesses. To help you understand the abilities of this tool and how it is constructed, I've divided it into six basic components:

1. The workpiece table
2. The router guide system
3. The fence with adjustable stop
4. The upper and lower guide stops
5. The router subbase
6. The router tray

The workpiece table—A panel router requires a flat, stable work surface with a straight edge for mounting the fence. I chose an ordinary 3-ft.-wide hollow core door for the table because it provides those things, and at $15, it cost less than what I could build it for. I mounted the table to a ledger on the wall. The ledger is 75 in. from the floor to give a comfortable working height. A 5-in. space from the wall gives enough clearance for the guide system. Standard door hinges let the table swing out of the way during storage, and side supports hold the table at a 65° angle when the table is in use.

The router guide system—Several years ago, I discovered that the aluminum extrusions used in Tru-Grip's Clamp 'N Tool Guides (manufactured by Griset Industries Inc.; see the sources of supply box on p. 29) interlock when one is inverted (see the photo at right). In this configuration, the two pieces slide smoothly back and forth with little side play, like a track. This system has several benefits: A panel can be set directly on the table without having to go under fixed guide rails. The guide is accurately located, and the panel is clamped tightly to the fence and to the table. The clamps are available in several lengths, but I've found that 36 in. is the most convenient (see the sources box). The manufacturer recommends using silicone spray to minimize wear.

The fence with adjustable stop—The fence holds the bottom edge of a panel straight, adds a runner for an adjustable stop and measuring system, and gives a place to mount the lower guide stop. Fence construction is partially dictated by the stop you use. I chose a Biesemeyer miter stop because it has two adjustable hairline pointers, which let you set and read both sides of a dado (see the top photo).

For the adjustable stop to work, the fence should be 1½ in. thick and the top edge of the fence has to be 1⅝ in. above the top of the table. My fence is two thicknesses of ¾-in. plywood laminated to form a 1½-in.-thick piece that is 3 in. wide and 96 in. long. To allow the router to pass through at the end of a cut, I made a 1-in.-deep notch in the fence. The notch is 13 in. long to fit my router. I located this notch 36 in. from the right, so I can dado in the center of an 8-ft.-long panel. To finish off the fence, I glued plastic laminate to the top, faces and ends. Before mounting, I cut a ¼-in. by ¼-in. groove in the back to provide for dust clearance, which ensures that the bottom of a panel stays flush to the fence. The fence is mounted to the bottom edge of the table with 2½-in.-long screws.

The upper and lower guide stops—The upper and lower guide stops allow the Clamp 'N Tool Guide to be set exactly

From *Fine Woodworking* (January 1995) 110:86-89

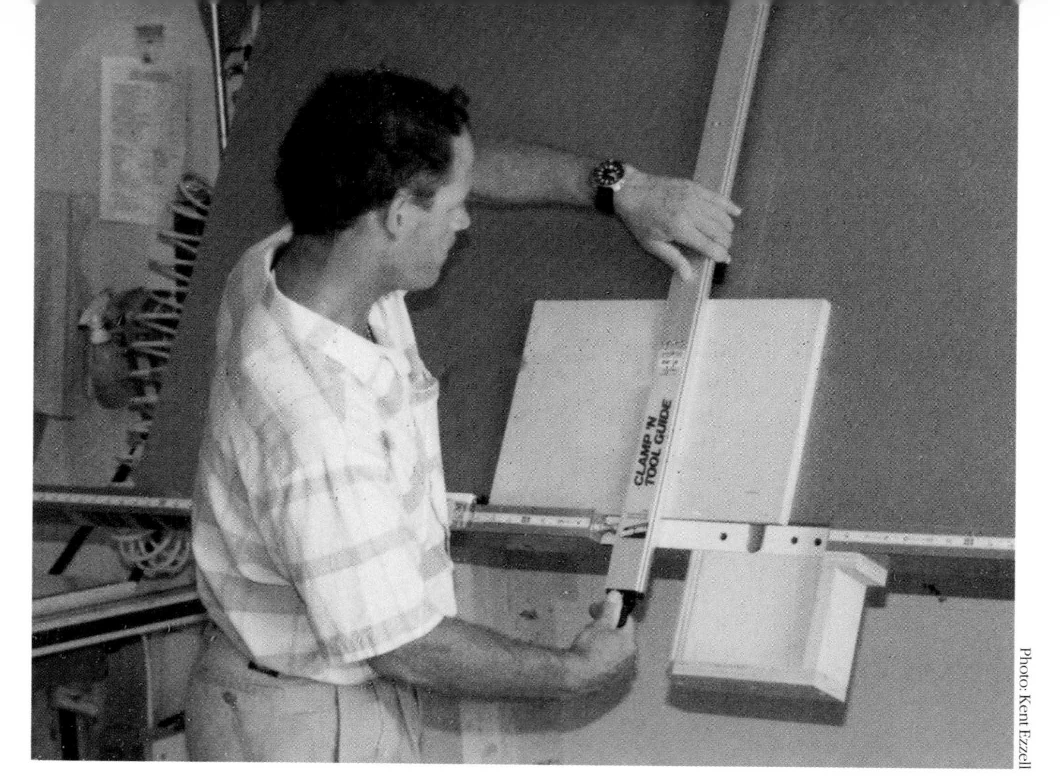

***Setup for dadoes is easy.** Just slide the Clamp 'N Tool Guide to the stops, and clamp the guide to the work by snugging up the black plastic dogs.*

Photo: Kent Ezzell

Commercial bits make clean cuts

Commercial panel routers work so well because the router bits are specifically designed to eliminate chipping and tearout, and they can also cut at higher feed rates. But their biggest benefit is that their cutter and arbor are two separate pieces (see the photo at right), which means that the arbor can stay secured in the router collet while you simply unscrew the cutter from the ½-in. arbor to change the bit size. Commercial panel-router bits (see the sources of supply box on the facing page) are available in a full range of sizes, including undersized ones for veneer plywood and oversized ones for two-sided melamine. An arbor and cutter set costs about $35, less than a decent-quality dado blade set.

When you need to change the width of a dado, select the correct cutter size, and screw it on the arbor (no wrenches required). The depth of cut doesn't need to be reset because the height of the cutter stays the same. This process is much quicker than using a dado blade on the tablesaw, where you have to use shims to get the proper width, and then make test cuts to set the depth of cut. *—S.L.*

***Panel-routing bits change easily.** The only things the author uses from industrial panel routers are the bits, which have interchangeable cutter tips.*

90° to the bottom edge of a panel. The lower guide stop is integrated in the fence (see the top photo on p. 27), and the upper guide stop is fixed to the top of the table. The lower stop is a ⅜-in. bolt threaded into a T-nut inset into a block and glued to a notch in the fence. The center of the bolt head should be 1⅛ in. above the work surface, or ½ in. above the bottom of the notch. The upper stop consists of two pieces of ¾-in.-thick plywood laminated to form a 1½-in.-thick piece, 12 in. long. The top is notched on both ends to leave a 2-in.- by 2½-in.-wide section in the center. Another bolt and T-nut are screwed to the shoulder. The center of this bolt is 1⅛ in. above the bottom of the notch. To fine-tune the stops for square, turn the bolts, and lock them with a nut. After the stops are set, adhere the measuring tape for the adjustable stops onto the top of the fence.

***The router subbase**—*Parts for the router subbase consist of a medium-density fiberboard (MDF) bottom, an upper base made out of ¾-in. plywood that mounts to the router, and a piece of upside-down extrusion screwed to the side so it can engage the guide track. Drawing detail B shows the dimensions I used to mount my Porter-Cable model 690 router. But you could modify the subbase to suit your router. Regardless of the router, the bottom should be ⅝ in. thick so that the extrusions interlock properly.

After the bottom is cut to size, center the baseplate on the bottom, and align the router handles at a right angle to the extrusion. Drill and countersink the mounting holes and mount the upper base to the bottom. Next, carefully, plunge a ¾-in. bit by slowly lowering the router motor. Then cut two dadoes, each ¼ in. deep by ¾ in. wide across the bottom. The first dado runs the full length and the second goes halfway across, 90° to the first. This T-shaped slot removes dust from the subbase (see the center photo on p. 27).

For the piece of inverted extrusion, I obtained stock from the manufacturer. But because they currently don't sell this separately, just buy a 24-in. clamp, and cut off the ends. I used a 21-in.-long piece.

The bottom of the router subbase slides directly on the face of the panel so that the depth of cut is registered from the top of the panel. This is desirable because when you switch material thickness from ⅝ in. to ¾ in., for example, the depth of cut does not have to be adjusted. Also, if the panel is slightly warped or some dust gets between the panel and the table, the cutting depth is not affected. Interchangeable bits also speed up the process (see the box at left).

***The router tray**—*The purpose of the router tray is to give the router a place to rest after it has completed a cut. The tray is mounted to the fence on the back side of the notched-out area. My tray is made out of ¾-in. plywood and is screwed to the fence. On the right edge of the tray, a piece of ⅛-in. Plexiglas protrudes into the tray opening. As the router slides down into the tray, the Plexiglas piece fits into a slot cut into the edge of the subbase and prevents the router from lifting out of the tray.

Using the panel router

The panel-router sequence to make a dado goes like this: First, I set the adjustable

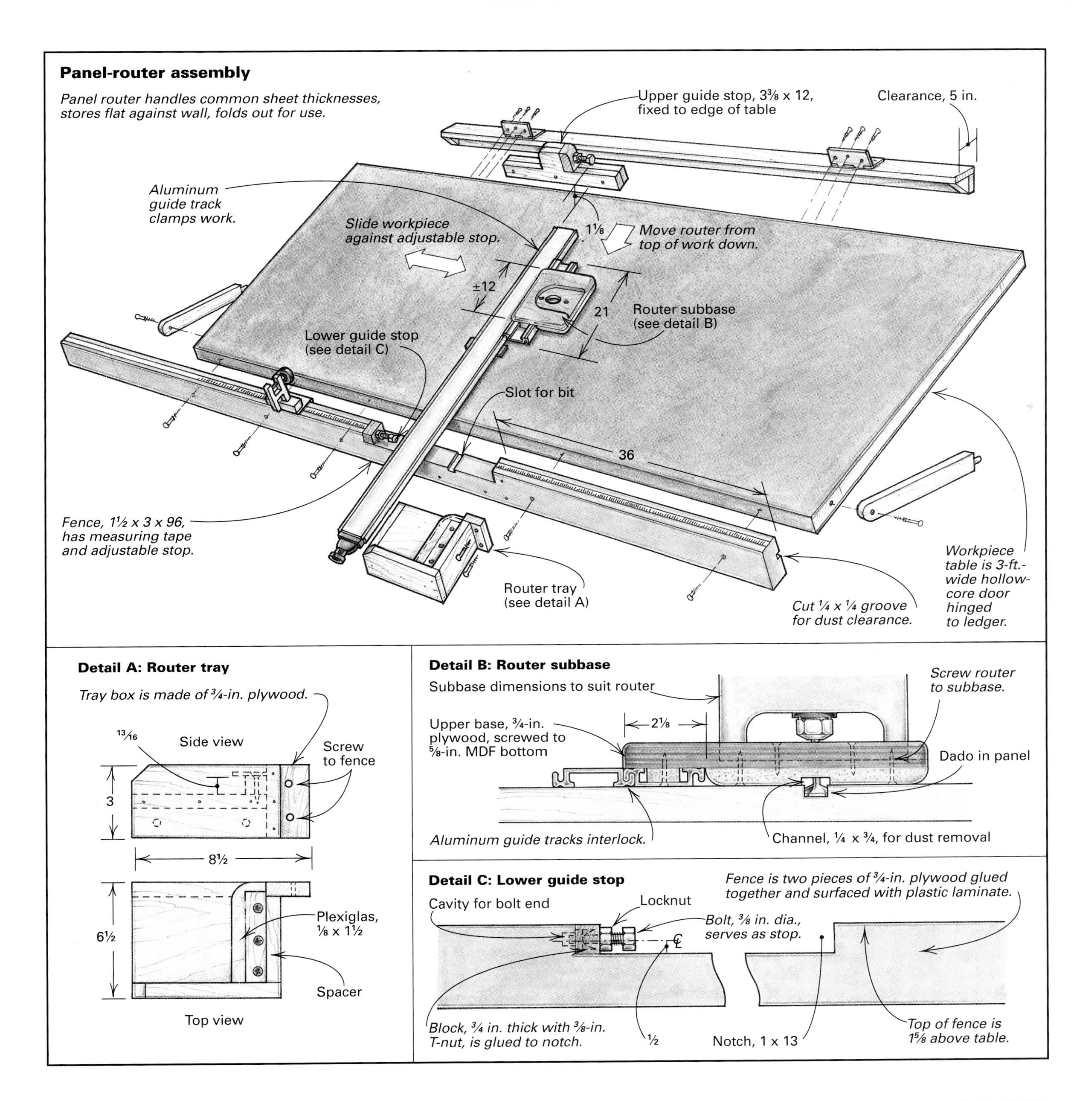

stop to locate the dado where I want it. Second, I set the panel on the table and slide it up against the adjustable stop. Third, I place the Clamp 'N Tool Guide on the panel, slide it against the upper and lower guide stops, and clamp it down (see the top photo on the facing page). In this one step, the guide is squared to the panel and clamped to the table. Fourth, I set the router on the panel with the extrusions interlocked. I hold the router subbase above the top of the panel so the bit clears. Finally, I turn the router on and cut the dado. To make stop dadoes, I insert a spacer block in the bottom of the tray to prevent the router from cutting all the way across a panel. While this setup may not be perfect for a large production shop, it is certainly affordable and conserves space. □

Skip Lauderbaugh is a sales representative for Blum hardware and a college woodworking instructor. His shop is in Costa Mesa, Calif.

Sources of supply

Clamp 'N Tool Guide
Griset Industries, Inc., P.O. Box 10114, Santa Ana, CA 92711; (800) 662-2892

Adjustable stop
Biesemeyer, 216 S. Alma School Road, Suite 3, Mesa, AZ 85210; (800) 782-1831

Panel-router bits
Safranek Enterprises, Inc., 4005 El Camino Real, Atascadero, CA 93442; (805) 466-1563

Drawing: David Dann

Shopmade Rip Fence Assembles Easily, Stays Aligned

Bolted steel components, setscrews and a toggle clamp are keys to accuracy

by Worth Barton

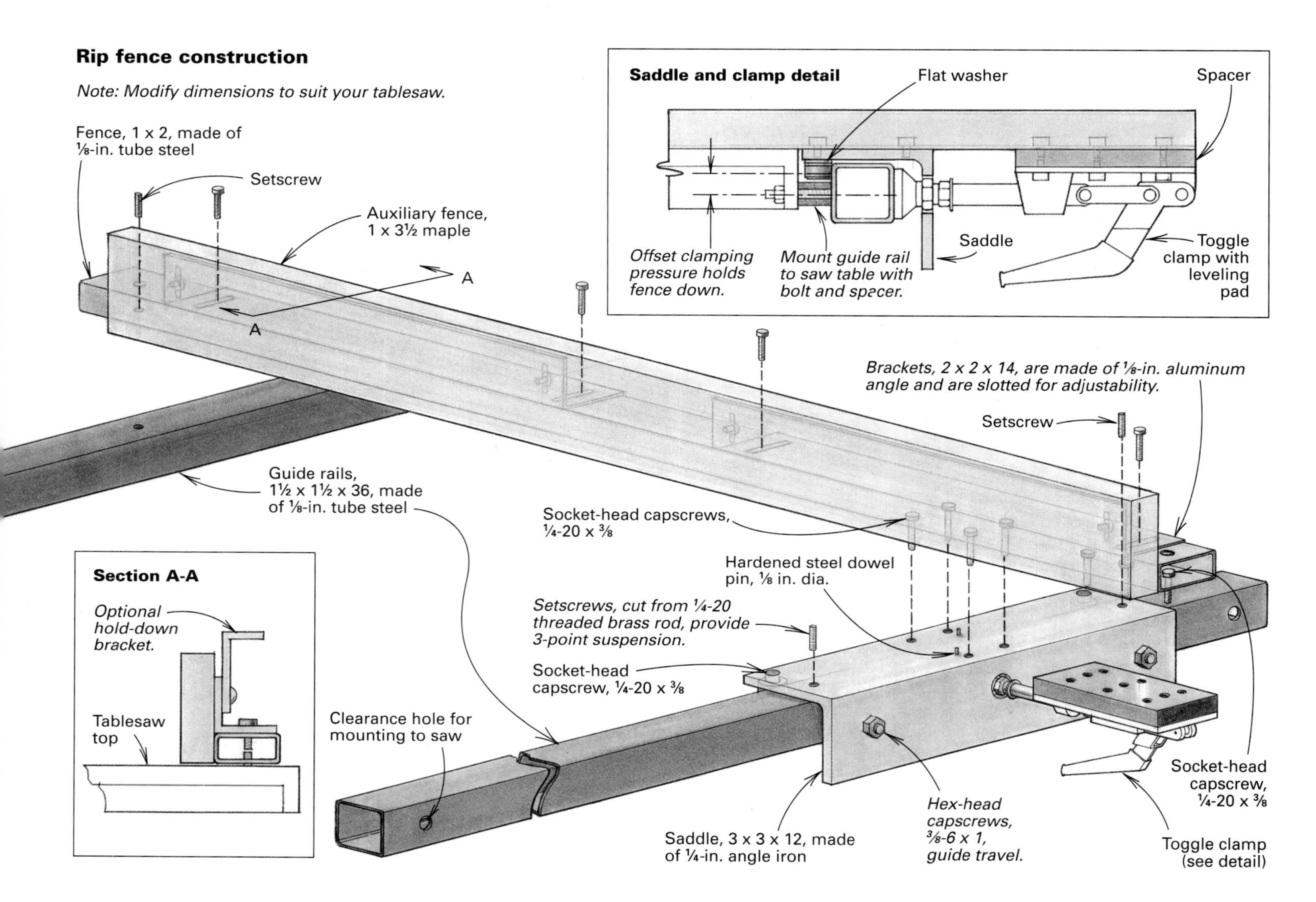

When I bought an older-model 10-in. Craftsman tablesaw, I was pleased with the saw's operation, but I was frustrated by its rip fence. It was a pitifully thin zinc diecasted saddle that soon broke. I could have bought one of the many after-market rip fences that are available (see *FWW* #68, p. 41). But I knew that I could make a sturdy, accurate rip fence fairly inexpensively (mine cost about $65) following a few simple ideas (see the top photo on the facing page).

Design

Building a first-rate replacement rip fence is pure fun—good for the shop and for the ego. I began by making a "got to have" list:

- Strength, durability and deflection resistance
- Reasonably available components
- Construction requiring only a drill press and hand tools
- Repeatable settings with low-friction movement
- Consistent clamping behavior
- Quick removal

Using square tube steel for the fence took care of the first three items, and a toggle clamp in the saddle satisfied the last three.

Steel parts plus single-rail locking equal precision—Most impressive of the commercial rip fences are the cast-iron and steel ones that marry precise surfaces to smooth movement. I decided

Photos: Alec Waters; drawings: Kathleen Rushton

Rip fence slides smoothly, locks positively*—Barton, setting up for a rip cut, snugs his fence in position using the saddle's toggle clamp. The fence, which has a hardwood auxiliary fence, was made out of standard steel sections and hardware. The fence slides on three brass setscrews that contact two square-tube guide rails.*

The underside of the saddle reveals how the fence stays aligned. *Two socket-head screws (left side of angle) and a toggle clamp (right side of angle) sandwich the front guide rail. The saddle is fixed squarely to the fence by bolts and steel dowel pins. The two other bolts (with nuts) guide the fence during positioning.*

to capitalize on those principles using common steel sections bolted together. I also decided to use a single-rail locking mechanism. Here's why: Many commercial rip fences, once positioned, get locked to both the front and rear guide rails. But when I checked a couple of fences of this type using a dial indicator, I found that the locking action would slightly skew the fence out of parallel. So on my rip fence, I made a saddle that has a toggle-clamp plunger offset below two guide bolts at the front rail (see the drawing detail on the facing page). This provides the fence with the necessary down pressure (to the table), which means that I don't need a clip at the rear rail. Though the rip fence still requires a rear rail, it is for guidance only—not for latching the rip fence in position.

The saddle, the piece that connects the fence to the front guide rail, is really the key. I chose a long base for the saddle to control what aircraft and boat designers call pitch, roll and yaw. To picture these phenomena, think of the fence as an airplane's fuselage. Pitch refers to the degree of nose-to-tail level. Roll is side-to-side (port to starboard) level. On the saw, these motions are relative to the table, the horizontal reference plane. To understand yaw, think of the fence's saddle as the tail of the airplane. The tail can swivel back and forth while remaining level with respect to the fuselage, as though the airplane was pivoting about a vertical axis. Yaw is similar on the saw, though it is greatly diminished; the rip fence can twist from side to side in the plane of the saw table, like a washing machine agitator.

To control yaw when the saddle position is fixed, I mounted two guide bolts (behind the front guide rail) and an adjustable De-Sta-Co toggle clamp between the two bolts (ahead of the front guide rail). The front rail is sandwiched between the ball-and-socket pad on the end of the clamp's plunger and the heads of the screws. That keeps the fence perpendicular to the rail. Because the screws in the saddle are above the plunger, clamping pressure forces the fence onto the saw table. This pressure enables the rip fence to resist the uplift action of a hold-down device. I use Shophelper hold-downs (available from Woodworker's Supply, 1108 N. Glenn Road, Casper, Wyo. 82601), which have anti-kickback rollers. Two

From *Fine Woodworking* (November 1994) 109:88-90

Brackets secure the auxiliary fence and a hold-down accessory. *The author used aluminum angles to attach the auxiliary fence to the main fence. The brackets are slotted to allow adjustment and to set minute tapers for ripping. Similarly, the add-on top bracket is for mounting the yellow-wheeled (Shophelper) hold-downs.*

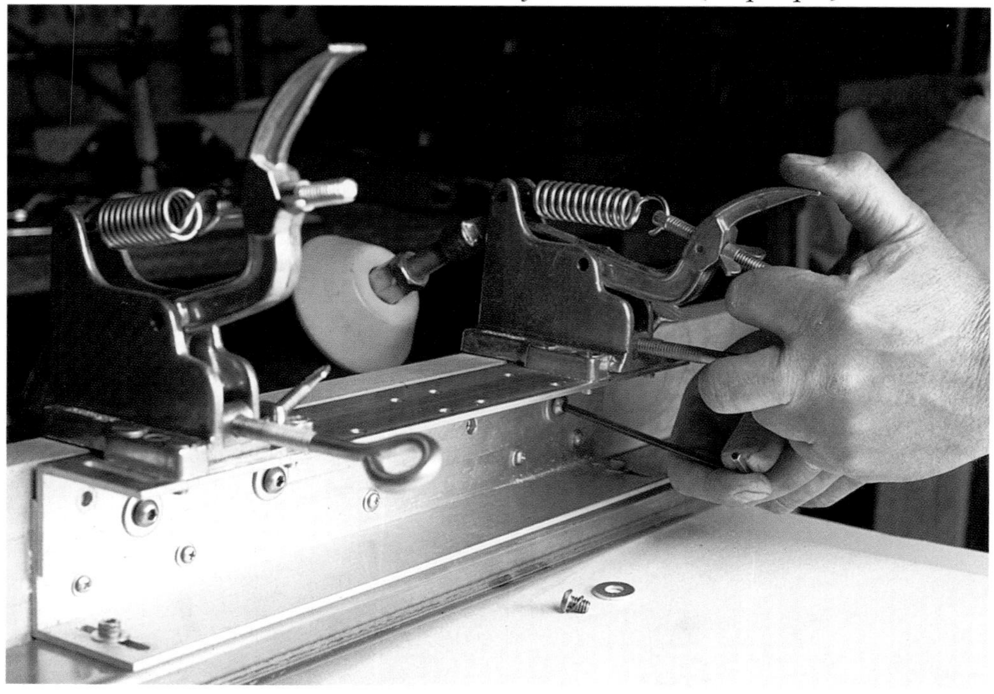

additional bolts, widely spaced and adjustable (see the bottom photo on p. 31), control yaw when I slide the saddle. This allows the motion to be smooth and free from lock-up.

Contact points: setscrews and guide rails—To restrict pitch and roll, I made the fence so it contacts the guide rails at three places: Two saddle points ride on the front rail (see the bottom photo on p. 31), and one point on the end of the fence rides along the rear rail. For the contact points, I installed three brass setscrews, which provide a means of leveling and act as low-friction bearings.

The guide rails must be square or rectangular because the saddle has to lock positively to the front guide rail, and the setscrews must slide on flat surfaces at both the front and rear guide rails. If your tablesaw has pipe rails or angle-iron rails, replace them with square-tube sections (see the drawing on p. 30). To attach the rails, drill through the flange (edge) of your saw table, so you can use bolts and spacers to hold the rails perpendicular to the miter slots and parallel to the table surface. Set the rails lower than the table, so the rails don't interfere with the miter slots.

Materials, fasteners and assembly

Because steel has three times the stiffness of aluminum, I used standard structural steel box shapes for the long members (the two guide rails and the fence). The tube steel is dimensionally uniform, has a high resistance to twisting and is readily available at most metal-supply houses (see sources of supply). I used 1/8-in.-thick wall tubing to avoid bolt tearout in the tapped holes. Have your steel vendor cut the tubes to the exact length you need. For the saddle, I used a 12-in. length of 1/4-in. by 3-in. by-3 in. angle iron. A model #607 De-Sta-Co toggle clamp locks the saddle; a piece of straight maple, attached by aluminum angles, serves as an auxiliary fence (see the drawing).

I bolted the parts together rather than welding them. Struggling with welding distortion can ruin your day. By contrast, bolts are easy to drill and tap for and are easy to remove. Suppliers offer a wondrous variety of fastening and clamping devices (see sources of supply). I use short fasteners because they reduce connection springiness and still afford some adjustability. Socket-head capscrews are ideal because they are made of high-quality steel and install easily.

The saddle-fence assembly is essentially a T-square, which glides on the front and rear rails. A .005-in. to .015-in. gap between the table and the fence allows clearance for sawdust and promotes smooth movement. As a safety feature, I extended the fence over the toggle clamp (see the top photo on p. 31), which prevents me from accidentally bumping the actuating lever.

Slotted brackets attach the auxiliary fence to the main fence (see the photo above). A similar bracket attaches the optional hold-down. The main brackets are symmetrical so that the auxiliary fence can be placed right or left of the fence. The slots, unlike holes, allow the fence to be adjustable and skewed for 2° or 3° tapers. It also enables the fence to be opened slightly at the rear of the saw (mine skews .020 in. over its length), as opposed to being parallel to the blade. A flared rip fence lessens the likelihood of kickback when you're ripping wet or unstable wood.

Setup for use

Make sure your sawblade and miter slots are parallel. Then set the fence parallel to the slots. To do this, place the assembly on the saw and attach a dial indicator to a miter-slot guide. Run the guide back and forth in the slot as you check the fence for runout. Tighten the screws joining the saddle to the fence. To ensure that the fence-saddle squareness won't be lost through rough handling, match-drill the parts so that you can press-fit hardened-steel dowel pins to lock the assembly: First clamp the saddle and fence together, and then drill and ream them to receive the pins. You can press in the pins with a drill press or tap them in with a hammer. □

Worth Barton is a design engineer, inventor and hobbyist woodworker living in San Jose, Calif.

Sources of supply

Clamp
De-Sta-Co, PO Box 2800, 250 Park St., Troy, MI 48007; (313) 589-2008. (Note: modify thread to 3/8-16 for stud-type plunger pad.)

Aluminum and steel
Adjustable Clamp Co., 417 N. Ashland Ave. Chicago, IL 60622; (312) 666-2723

Castle Metals, 3400 N. Wolf Road, Franklin Park, IL 60131; (708) 455-7111

Tooling accessories and fasteners
Reid Tool Supply Co., 2265 Black Creek Road, Muskegon, MI 49444-2684; (800) 253-0421

Vlier Corp., 2333 Valley St., Burbank, CA 91505; (818) 843-1922

Make Your Own Dovetail Jig

Quick and easy system for routing this traditional joint

by William H. Page

The blanket chest I wanted to make for a gift was basically a large box joined with dovetails at the corners. I didn't have enough time to hand-cut the joints, and I didn't want to pay $300 for a commercial jig to do the job, so I set to work developing my own jig.

Shop-built from scraps, these unusual jigs, one for the tails and one for the pins, cut tight-fitting through-dovetails (see the photo below), a task that even many commercial jigs can't handle. Designed for routing dovetails for large carcase construction, the jigs can be built in less than two hours for just pennies.

Layout is quite simple and can be done as the tail jig is being assembled. Fingers screwed to the tail jig guide the router bits; the key is ball bearings. The bits used to cut the joint are guided by bearings the same diameter as the cutter. Pin and tail size and spacing are variable, and jigs can be built to handle any width board.

Basics of jig construction

Before making any of the jigs, the project stock must be jointed, planed and cut to final dimensions. The stock should be flat and square, and be sure to include a couple extra feet of stock for making and testing the jigs. The jigs are assembled around some scraps cut from the actual stock. This way, the jigs precisely fit the stock and eliminate the need to fiddle with adjustments or set-up routines, ensuring perfect-fitting dovetails.

I start with the tail jig, and in the process of making this jig, I also cut a guide board that precisely locates the pin templates for as-

***Precise through-dovetail joints** (see inset photo) are easy to rout with the aid of a couple of shop-built jigs. Here, the author completes the second part of the joint by routing the pins with a bearing-guided straight bit. The bearing rides against pin templates that have been positioned accurately using a guide board routed with the tail jig, which is the first jig to be built.*

Photos: Charley Robinson

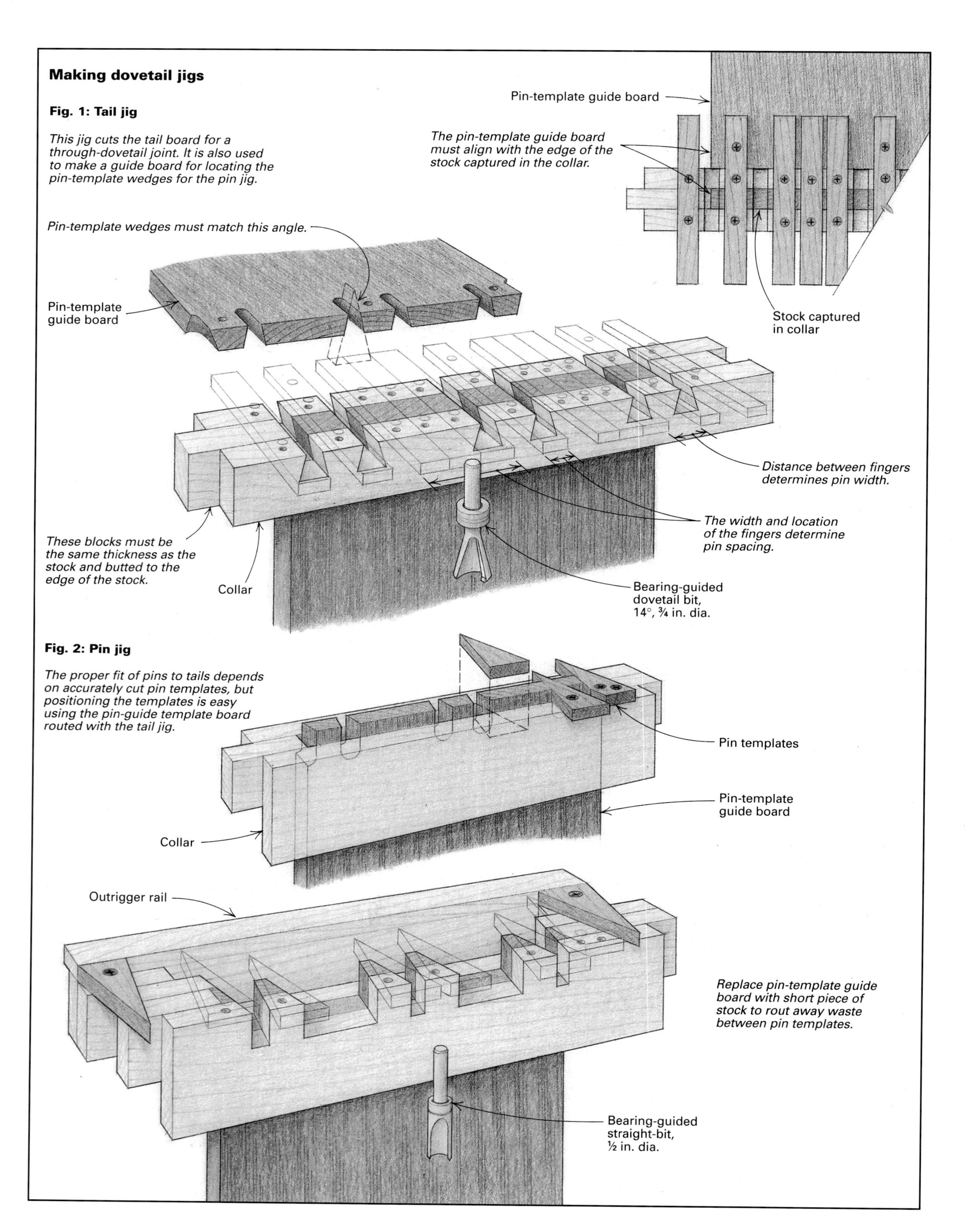

Drawing: Vince Babak

sembling the pin jig. Using the tail jig to rout the pin-template guide board ensures a perfect match of pins to tails.

The tail jig consists of a collar that surrounds the stock to be joined and a series of fingers screwed to the top of this collar, as shown in figure 1 on the facing page. The fingers serve as a stop when inserting stock into the collar and as a guide for the bearing on the bit. The location of these fingers across the top of the collar determines the spacing of the pins.

With the fingers in place, I run the dovetail bit through the collar of the jig and a scrap piece of stock clamped in the jig. These cuts create the tail piece, or the openings that the pins will fit into, and prepare the jig for use. The collar must be clamped to the stock to avoid any movement that could affect the accuracy of the cut.

The pin jig consists of a collar built around the pin-template guide board, but instead of straight fingers, the pin templates for this jig are wedges with an included angle to match the cut of the dovetail bit, as shown in figure 2 on the facing page. An outrigger attached to the pin collar provides full support for the router when routing the pins.

With both jigs assembled, I rout a joint in a couple of pieces of scrap clamped firmly in the collars to test the fit and to be sure I like the pattern before proceeding with my good stock.

Making the tail jig

To make the tail jig (see figure 1 on the facing page), I clamp a short piece of the prepared stock in my bench vise. I begin by building the collar around this piece of stock, using 2-in.- to 3-in.-wide scraps that are about 4 in. longer than the width of the stock. The collar pieces are clamped flush to the end of the stock so that they overhang equally on both sides of the stock. The end collar blocks should be butted tightly to the side of the stock, as shown in figure 1.

The guide fingers that are glued and screwed to the top of the collar are simply strips of hardwood or plywood about ⅝ in. thick and about 8 in. long. Position the strips for any pin pattern that you want, but keep in mind that the pins must be at least ¾ in. wide, the diameter of the bearing that will ride against the fingers. Also, the distance between the pins must be at least equal to the diameter of the straight bit used to cut the pins. The fingers also must be square to the collar. To avoid pin cutouts where I don't want them, I fill gaps between the fingers.

To rout the tail jig, I chuck my bearing-equipped dovetail bit in the router and set the depth of cut about 1/32 in. deeper than the thickness of my prepared stock. (Bearing guided bits can be made by gluing a bearing the same size as the bit to the bit's shaft, or they can be ordered from Freud, 218 Feld Ave., High Point, N.C. 27264.) I then rout the tails by running the bearing between the fingers. I make two passes in each slot to be sure the bearing rides firmly against both fingers for each pin cutout or else the pins won't align properly with the tails. This completes the tail jig, and in the process, I've made a scrap tail piece to test the fit of the joint.

To make a tail jig, *build a holding fixture that forms a collar around the stock, screw guide fingers to the top edge and then rout between the fingers to create the sockets.*

Making the pin jig

The first step in making the pin jig is to use the tail jig for cutting the guide that locates the pin templates, so the pins and tails line up. I do this by butting a piece of prepared stock against the back side of the tail collar and screwing down through the fingers and into the jig stock. This piece of stock must be the same width as the workpieces to be joined, and its edges must align with the edges of the jig, as shown in figure 1 on the facing page. To create the pin guide, I run my router between the fingers of the tail jig as before, cutting approximately an inch into the stock, as shown in the photo below. After routing, I unscrew the pin guide and then clamp it in my bench vise with the routed end up.

As with the tail jig, I build a four-piece collar around the pin guide clamped in the vise. I let the pin guide extend about ½ in. above the collar, so the routed slots can be used to position the pin templates on the collar.

The pin templates are ⅝-in.-thick wedges that I cut on the tablesaw from a long strip about 3 in. wide. I set my miter gauge to 14°(because I used a 14° dovetail bit), cut one edge of the wedge, flip the strip over and then cut the other edge of the wedge. I test-fit the wedge into the pin guide, make minor adjustments to the miter gauge as necessary and then cut a new wedge. I continue this process until I get a wedge that fits snugly into the pin guide with no gaps on either edge (see figure 1). Then I cut a wedge, or pin template, for each slot in the pin guide.

I then push the pin templates firmly into place on the pin guide and glue and screw the templates to the collar. To fully support the router, I needed to attach an outrigger rail to the collar in front of the pointed end of the templates.

To route the pins, I set up a second router with a bearing-guided, ½-in.-dia. straight bit. Again, the depth of cut is just a hair deeper than the thickness of the stock. Before routing away the waste between the pins, I removed the pin-template guide board from the jig and replaced it with a short piece of the prepared stock. Then, with the router sitting on the pin templates and the outrigger rail, I routed away all the material on the collars and the scrap stock that is not covered by the pin templates, as shown in the photo on p. 33. I used firm pressure to be sure the bearing rode tightly against the templates for an accurate cut. Routing the waste completes the jig and cuts a pin test piece.

I remove the test piece and try it in the tails previously cut. If the joint is too loose or too tight, it's usually a result of not keeping the guide bearing firmly against the sides of the fingers or pin templates. You might want to try running the router through the jigs again with new test pieces in place. Minor misfits can be adjusted by shaving the edge of the pin templates or adding masking-tape shims. If you're satisfied with the fit and spacing, slide the appropriate jig over the end of your stock and start cutting. The actual routing of joints takes about five minutes each. □

Bill Page is a woodworker in Toledo, Ohio.

From *Fine Woodworking* (May 1994) 106:79-81

***Clamping up a large carcase** is much easier with the author's carcase-press clamping system than with ordinary pipe or bar clamps. The press consists of two units, each of which is made of four veneer-press screws, a couple of lengths of heavy metal strapping, a few board feet of hardwood and a handful of nuts, bolts and washers.*

Dovetailing Large Carcases

Dedicated bench and clamping system simplify and square the work

by Charles Durham Jr.

I made my first dovetailed carcase with wide pine boards salvaged from the original kitchen in my first house. Dry, flat and wide, those boards became a wonderful blanket chest. Since then, much of the lumber I've used on large-carcase projects has been less than ideal. Wide, flat and dry are more the exceptions than the rule, whether you use naturally wide boards or glue narrower stock to width. When wide boards are cupped, twisted or both—even a little—making dovetails that fit well is tough. Yet accurately fitted and squared dovetail corners are crucial to the success of large projects like blanket chests, highboy tops and slant-front desks.

The other problem with large-carcase projects is the glue-up. Even if you've cut good, accurate dovetails, gluing and clamping big boards can be a real headache or, worse, result in a flawed project—especially if you work alone, as I usually do. Having the pipe clamp I just tightened fall off and dent the carcase as I tighten the next clamp is just one more hassle than I need.

I solved both problems by building two assemblies: a dovetailer's bench to hold the boards flat, secure and indexed for accurate layout and cutting (see the drawing on the facing page and the photos on p. 38) and a carcase-press clamping system to help me close wide joints with uniform pressure, without having to wrestle an armload of clamps (see the photo above). Material for both is available at any good lumberyard, and you'll find all the hardware you need either at your local hardware store or through mail order. Total cost for materials was about $300, with lumber being the most expensive item. By substituting construction lumber for the hard maple I used, you could halve that amount.

Dovetailer's bench

The problem with laying out and cutting dovetails on a typical cabinetmaker's bench is that most benches are about 32 in. off the floor, which constrains you to narrower carcase work. To do bigger jobs on an ordinary bench, you have to jury-rig a support and clamp system to hold things flat and steady at the right height while you mark, saw and chop. My bench is a large, elevated clamping device that lets me overcome warp on wide boards, allowing me to dovetail the largest boards with ease and precision. The bench's working surface is at elbow height: 42 in. off the floor, which is long enough for the longest pin member I'm likely to encounter.

Photos: Vincent Laurence

Dovetailer's bench

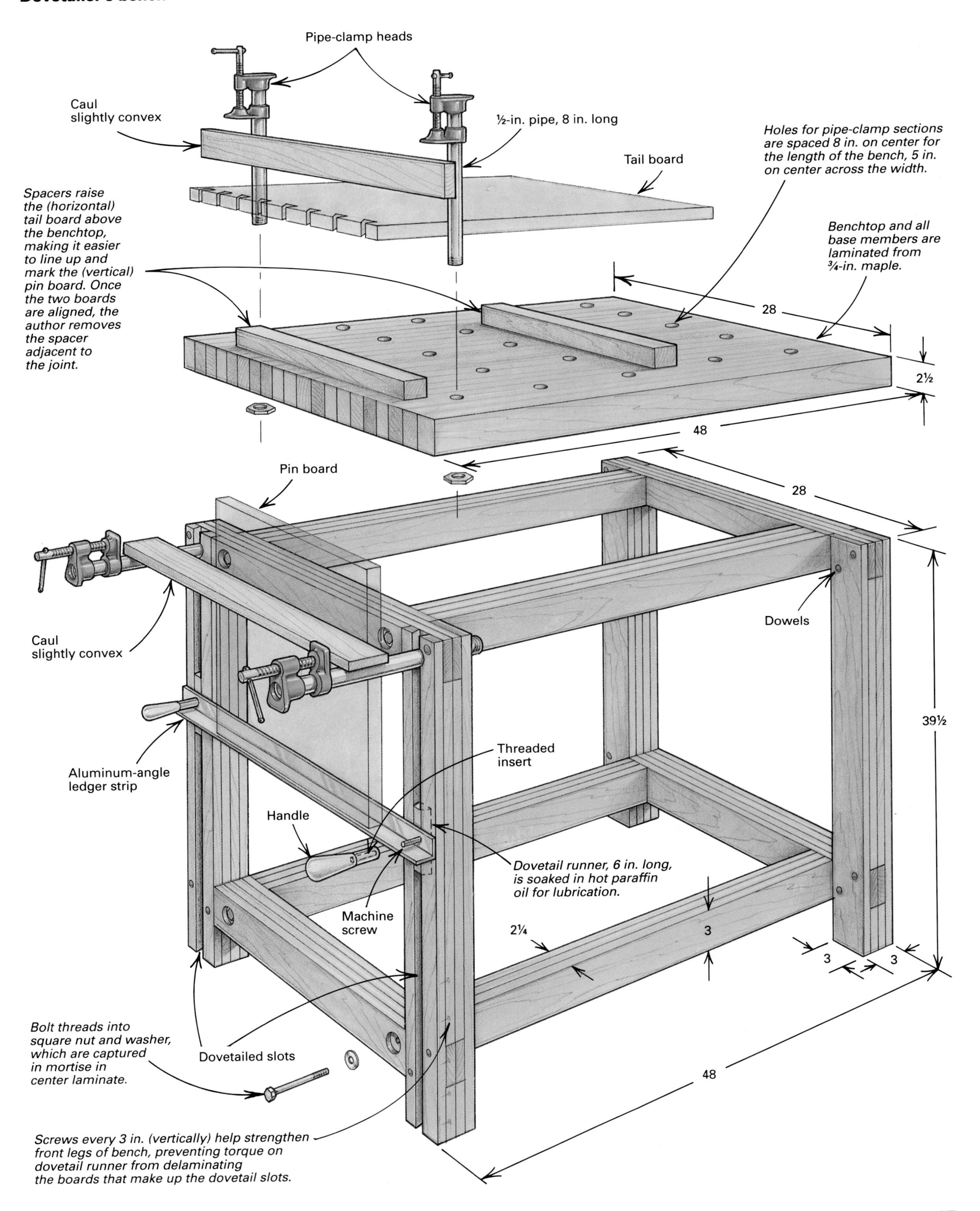

Drawing: David Dann

The deepest carcase I would ever dovetail is about 25 in. So I added space for the clamp heads (see the drawing) to establish the benchtop's width of 28 in. A 72-in.-wide breakfront was the longest project on which I saw myself using the bench, so I decided to make it a bit more than half that length (48 in.) to keep that breakfront's top and bottom from falling off.

I use pipe-clamp heads to hold boards in place (see the top photo on this page) and cauls extending across the bench's width to take out any warp in either board. An aluminum angle that raises, lowers and locks with a twist of the wooden handles serves as a ledger strip for the pin member (see the drawing on p. 37).

I cut dovetails in a fairly conventional manner, but with a couple of twists. I lay out the tails first, using a sheet aluminum template I made for the purpose. Then I saw to the line with a Bosch barrel-grip jigsaw and chop the waste out on my dovetailer's bench. The jigsaw is so much faster and is at least as accurate (probably more so) as cutting with a backsaw. I mark the pins from the tails, aligning the tail board on the benchtop with the pin board on the aluminum ledger, using a chisel and mallet to transfer lines (see the bottom photo on this page). A light, clean rap ensures a sharp line with no chance of following the grain, which can happen when marking with a knife. Again, I use the jigsaw, this time with its base set at approximately 14° (from a bevel-square set on the tail board) to cut to the line and then chop out the waste on the bench. The fit I get with this system is nearly perfect.

Marking tails*—The author uses an aluminum template to mark out the tails on the side board of what will be a mahogany blanket chest. The short sections of pipe clamp at the front of the dovetailer's bench ensure the board remains flat for an accurate layout.*

Marking pins from tails *is more certain with a chisel than with a knife because there's no danger of the chisel following the grain. It's important, though, to make sure the chisel is absolutely perpendicular to the surface of the board you're dovetailing.*

Carcase press

My carcase press will close any size project I'll ever build and will do it in much less time than it takes with loose clamps. With the time saved, I can close the joints correctly before the glue grabs. The only fixed dimension is its internal working height—enough to take those 25-in. boards I produced on the bench. The carcase press consists of a pair of clamping frames made of maple laminations and pre-punched, galvanized steel strapping. The head member of each is fixed and has veneer-press screws mounted to it. (Veneer-press screws are available from Constantine, 2050 Eastchester Road, Bronx, N.Y. 10461; 800-223-8087.) A foot member moves along the galvanized strapping to accommodate carcases of various widths. The clamping frames can themselves be positioned as near or far from one another as need be (see the photo on p. 36).

At each end of the maple laminations, I made a sawcut precisely as deep as the strapping is wide and drilled holes for the bolts that connect the wooden end pieces to the metal strapping. The straps I use are 60 in. long, but they're available in virtually any length. Smaller wooden cauls ride on the strapping to transfer the clamping force from the press screws to the carcase. Ideally, the clamping force should bear directly on the corner of the carcase, but I find that placing the force just inside the joint, right on the baseline, works just as well. With the 8-in. press screws and this setup, there's a range of about 4 in., fully opened to fully closed.

The elimination of loose clamps is the major benefit provided by the carcase press. Instead of watching and worrying about clamps falling off, I can monitor the joint. But there's another advantage. Quite often, clamping a project together forces it out of square in one plane or another. With loose clamps, the unending adjustment required to restore squareness can be maddening. None of that has been necessary since I began using the carcase press.

Moreover, when using loose clamps, if a carcase winds (so that diagonal corners are high), there's nothing you can do with ordinary clamps. With the carcase press, I just wedge shims between press and carcase in the high corners, and it's flat again.

In using the carcase press, I work at table height on a sheet of laminate-covered particleboard. Because the bottoms of both clamping frames that make up the press are square, they stand upright on their own, making it easy to slide the carcase into the press. I get the joints just started outside the press and then place it inside and dry-assemble the carcase. Only after checking to see that everything's going to close up properly do I apply glue and clamp the carcase for good. □

Charles Durham is a professional woodworker in San Clemente, Calif.

From *Fine Woodworking* (March 1994) 105:50-52

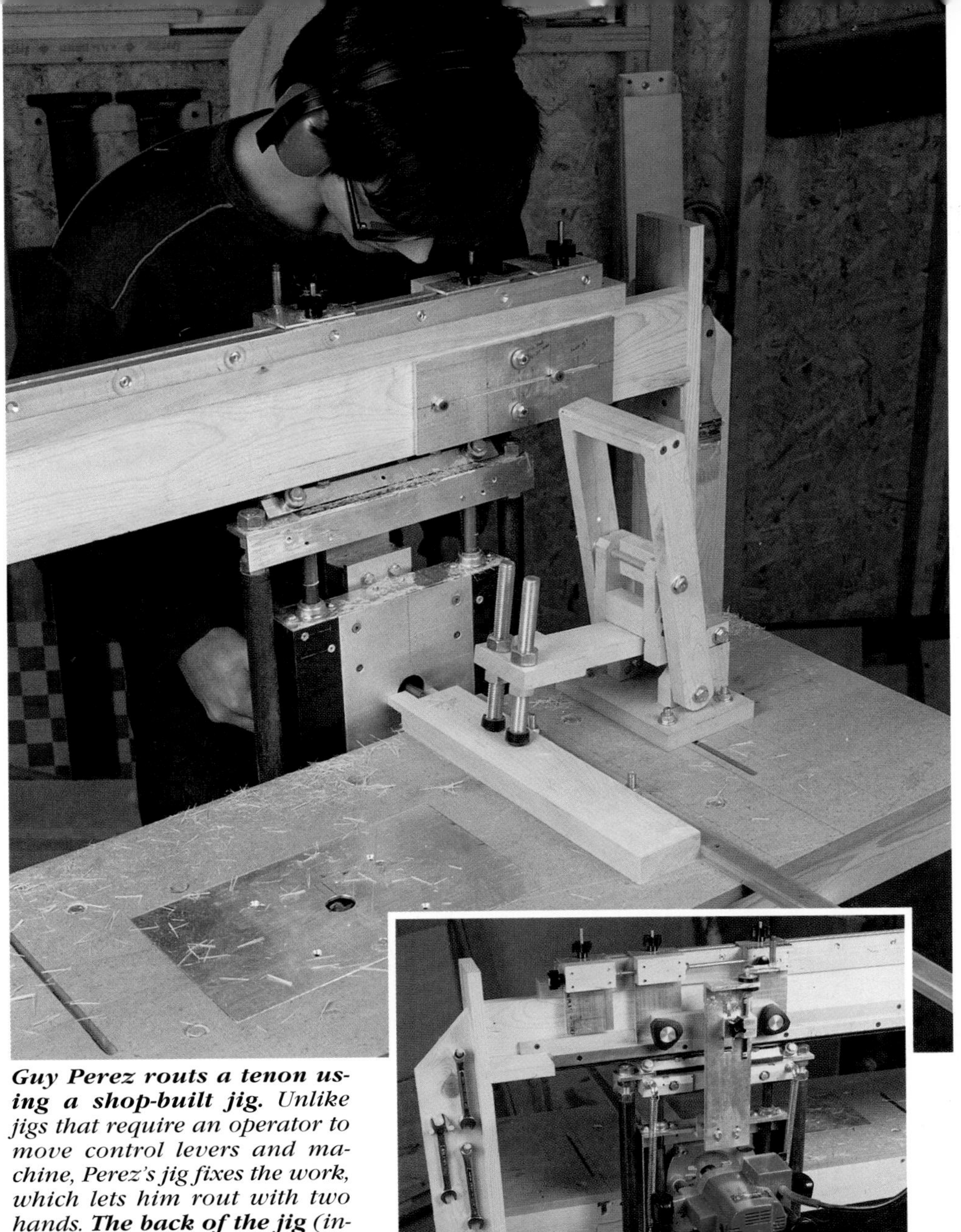

Guy Perez routs a tenon using a shop-built jig. *Unlike jigs that require an operator to move control levers and machine, Perez's jig fixes the work, which lets him rout with two hands.* ***The back of the jig*** *(inset) reveals a carriage that guides both vertical travel (rods through bushings) and horizontal motion (rails captured by bearings). The router's plunge mechanism sets depth of cut.*

Turn a Router into a Joint-Making Machine

Jig and templates tackle involved joinery

by Guy Perez

I often turn to my router for joinery tasks. With a fence and straight bit, the router makes quick work of mortise-and-tenon joints. And for most projects, the consequent problem of either rounding the tenon or squaring the mortise is relatively minor. However, I recently made a crib with 44 slats that required 88 mortise-and-tenon joints. This daunting task prompted me to build my own version of the fancy joint-making machines I had often admired but had never been able to afford. I based the jig around my joinery needs and the funds I had to work with. And I confess that I arrived at much of the design during some less-than-inspirational philosophy seminars.

The jig I built operates somewhat like the commercially available machines, such as the Matchmaker or the Multi-Router, which cost from around $600 to over $1,500. With my jig, the work stays fixed, which means I can move the router with two hands (see the top photo at left) instead of having to manipulate the workpiece, router and control levers. This makes the jig especially useful at routing the edges of large stock because it's much easier to move the cutter past a piece rather than the other way around. I can also position a template quickly and accurately relative to the stock, which eliminates much of the trial and error that's required to set up some joint makers.

In addition, the jig's templates are mounted above the router (see the inset photo at left), which makes them easy to see and keeps them away from the dust. Finally, the templates interchange quickly, and their holders easily adapt to different joinery, such as mortises and tenons, finger joints and dovetails.

Constructing the jig

Although it may look complicated, my joint-making jig was fairly easy and inexpensive to build (around $160, depending on the amount of work you have done by a machine shop). Basically, the jig is a plunge router mounted horizontally in an upright, linear-motion (X-Y) carriage, which is secured to a frame and table. A following device copies patterns secured by a template holder. I simply clamp the stock to the table, and trace the template with the follower as the router cuts the joint.

As shown in the drawing on p. 40, the jig has five subassemblies: an X-Y carriage, a wooden frame that has a platform and a table, a horizontally adjustable template holder, a vertically adjustable follower, and a fence and hold-down to position and clamp stock. I sized the frame to suit

Photos except where noted: Alec Waters

my joinery needs, and then I built the rest of the jig around this.

To construct the carriage, you could buy the aluminum bar and flat stock from a metal supplier, but I picked up scrap aluminum for under 70 cents per pound. I cut all the aluminum pieces to length on my tablesaw fitted with a carbide blade. Then knowing that I needed a few large holes in the pieces that I couldn't bore with my hand-held drill, I took the aluminum to a machine-shop equipped with a CNC mill/drill. The shop performed the work for only $30. Shops with conventional equipment gave me quotes around $100.

Carriage—The X-Y carriage consists of two major components: a vertical router carrier and a horizontal roller assembly. The router carrier holds the router and provides up-and-down movement by means of four bronze bushings that ride on two ⅝-in.-dia. steel guide bars. Not expecting ever to have to rout more than 2-in.-thick tenons or dovetails, I allowed just 3½ in. of vertical travel. I mounted the bronze bushings in self-aligning pillow blocks made of stamped steel. The blocks are available from Northern Hydraulics, P.O. Box 1499, Burnsville, Minn. 55337; (800) 533-5545; or you could use linear-motion bearings, which are carried at most bearing-supply shops. The vertical bars are fastened to the horizontal roller assembly, which relies on four pairs of precision roller-skate bearings for motion. The bearings are bolted to ¾-in. aluminum-angle brackets. These bearing brackets are fastened to 1½-in. aluminum angles so that the bearings are oriented 45° on either side of two horizontal steel rails (see drawing detail A). The bearings and rails work similarly to the guide system I used in a sliding saw table (see pp. 13-15). I used 41-in.-long rails, which allow for 28 in. of horizontal travel.

The router carrier is made from four lengths of 1½-in. aluminum angle riveted together at the corners. I used thicker, ¼-in. angle for the upper and lower pieces of the carrier because they support the bronze bushings. A ⅛-in.-thick aluminum plate mounted between the angles serves as a base for the router and stiffens the assembly. To keep the overall size down, I made the router carrier as small as possible, leaving just enough room for the router base to fit easily between the guide bars. Because the alignment of the bronze bushings is critical, I had their clearance and mounting holes professionally machined.

I connected the bearing-bracket assemblies with two ⅝-in.-dia. by 18-in.-long

Router jig assembly

The router travels via linear-motion guides. A stylus traces the pattern while the router's plunge mechanism controls depth of cut. With workpiece clamped to table, operator stands on platform and moves router with two hands.

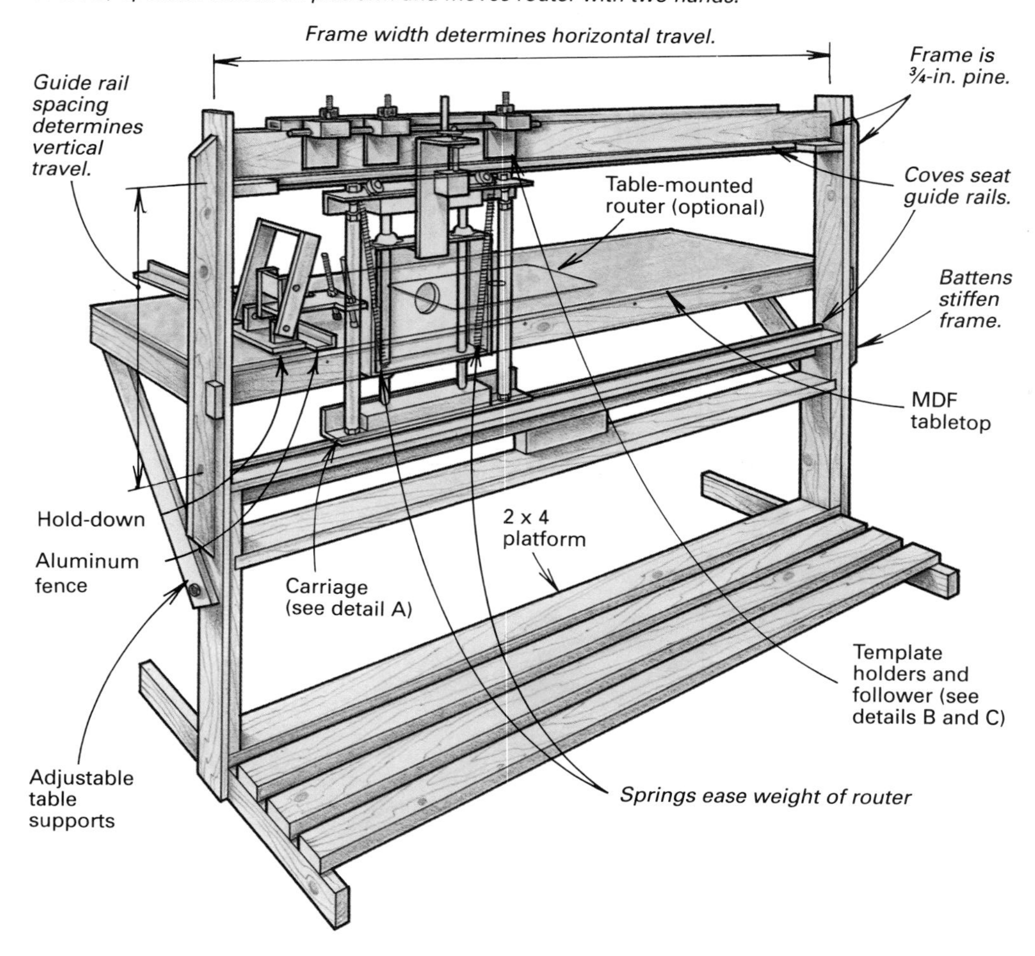

Detail A: X-Y carriage

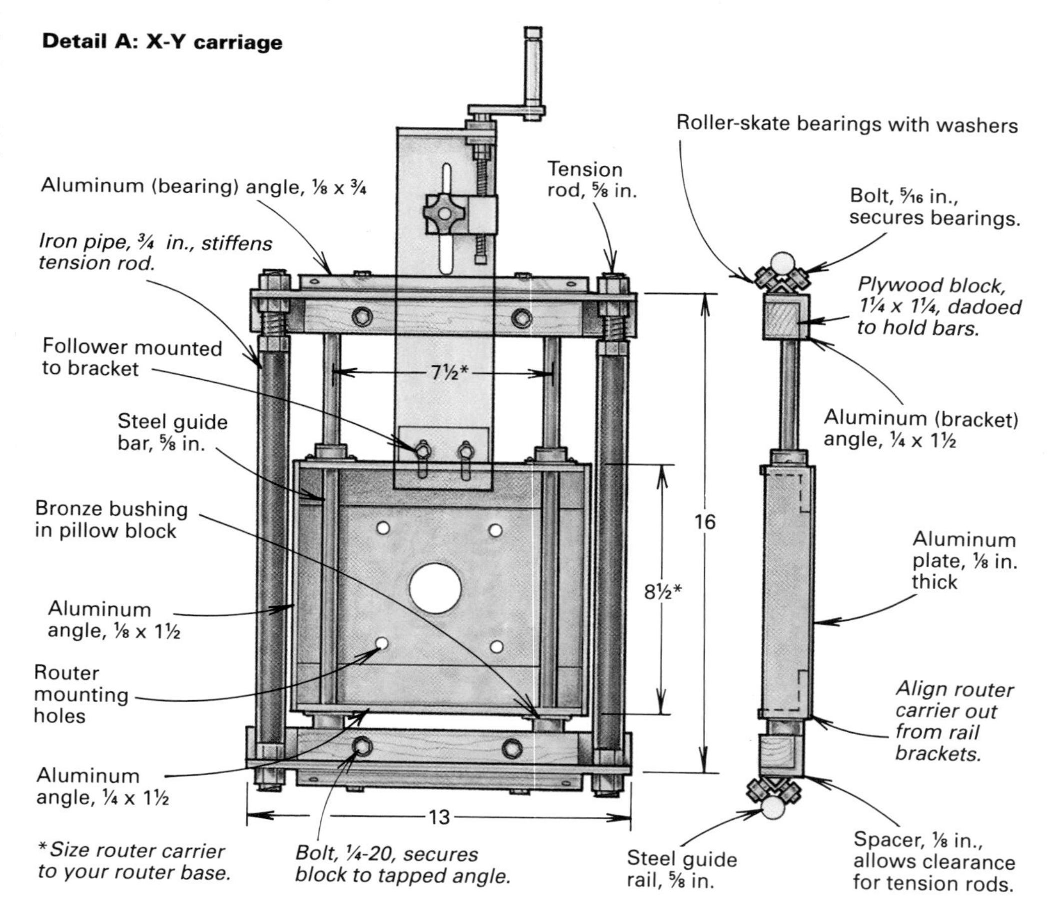

From *Fine Woodworking* (November 1993) 103:97-101

***A pattern to follow**—With a template bolted in place, the jig is ready to rout dovetail pins. The template holders (top) are adjustable left and right. The template follower (right) is height adjustable.*

threaded steel tension rods. The tension rods are stiffened by slightly shorter lengths of ¾-in. iron pipe. I had to cut some of the aluminum angle away so that the inside upper tension-rod nuts could turn freely. This allows just enough room to adjust the carrier for a tight fit to the guide rails. I secured the vertical guide bars to the horizontal brackets with two ¾-in. plywood mounting plates. I cut the bar-aligning dadoes from a single piece and ripped the two mounting plates from it.

***Frame and table**—*After assembling and mounting the horizontal bearing brackets to the router carrier, I set the tension rods to allow for ½ in. of adjustment either way. Holding the guide rails in place, I measured the outside distance to determine the inside height of the frame. I subtracted ¼ in. from the rail's out-to-out dimension to allow for cove cuts in the frame for the rails. I located the coves so the front of the carrier rides proud of the frame to provide clearance for machining longer stock. The frame width is determined by the length of the horizontal guide rails.

I initially built the frame from ¾-in. pine and later added battens to stiffen the frame. I think a 6/4 hardwood frame would be better. Also, I soon discovered that the frame provides a ledge for chips to build up on, so if I were to build the jig again, I would turn the frame boards on edge and cut bevels on either side of the lower guide rail.

The table is made of pine with a medium-density fiberboard (MDF) top, which can be slid away from the carriage to allow clearance. I also cut holes and slots in the table, so I could mount an aluminum fence and a shopmade hold-down (see the box on p. 43). In the extra table space, I made a cutout for a vertical router (see the photo on p. 39). The lower braces support the table and keep it square to the carriage.

***Template holder and follower**—*When I designed my machine, I was concerned with providing a way to hold the template and allowing crude lateral adjustments. And I knew that the follower should be rigid and height-adjustable. For setup, I initially relied on a cut and nudge method: Take a trial cut, estimate the error and nudge either the template or follower to compensate. But it didn't take long to pro-

Detail B: Template holder

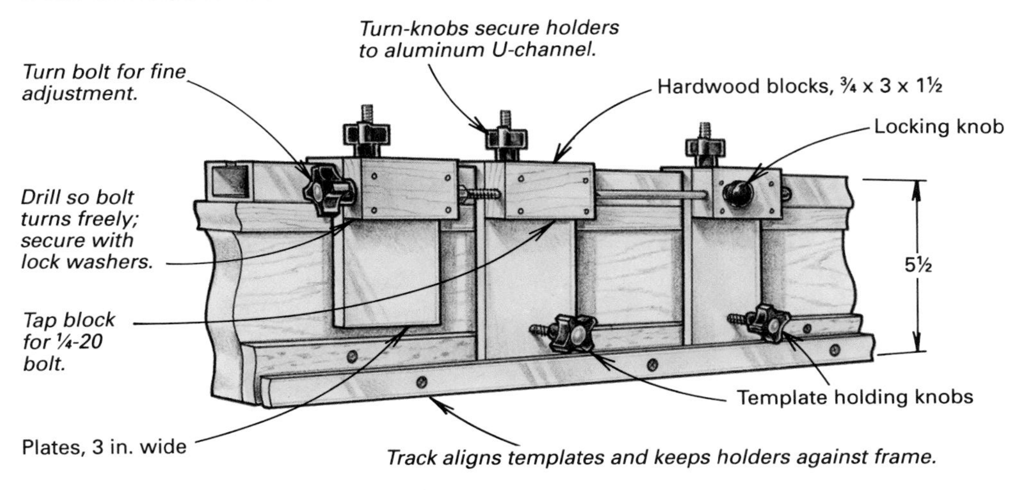

Detail C: Template follower

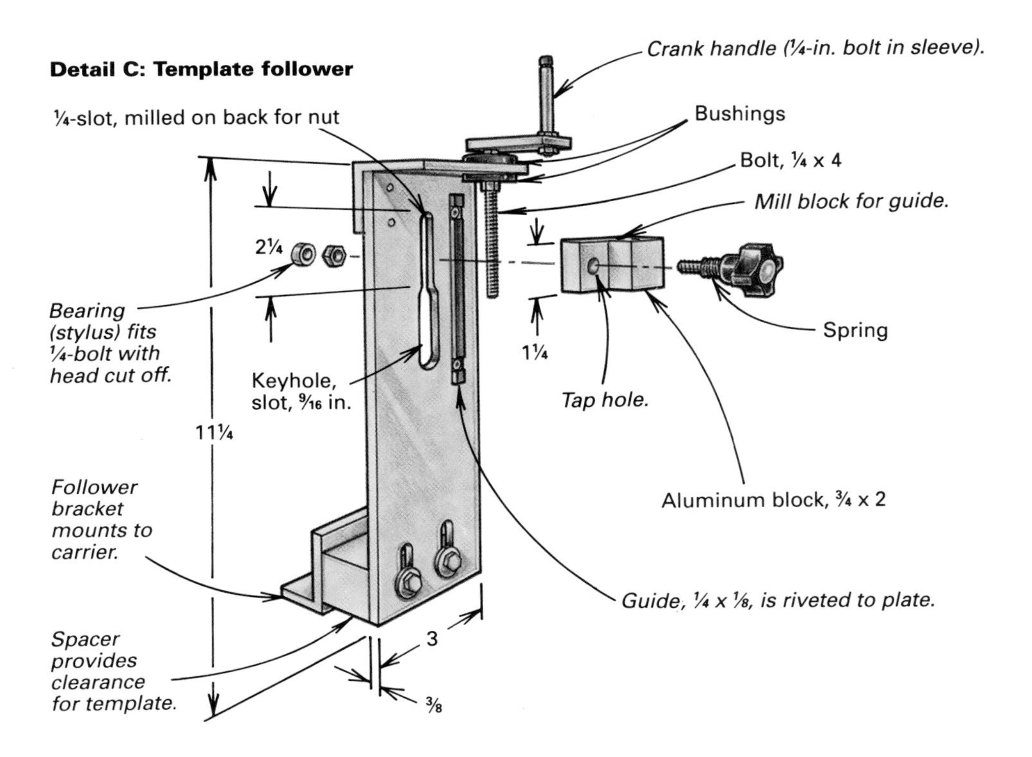

Drawings: David Dann

duce a pile of waste-tenons that way.

To remedy this situation, I introduced screw-driven adjusting mechanisms into both the template holder and the follower (see drawing details B and C on p. 41). The template holders consist of three brackets constructed from ¼-in. aluminum plate and 1½-in. angle. The brackets can be individually locked to a guide track by turnknobs (available from The Woodworkers' Store, 21801 Industrial Blvd., Rogers, Minn. 55374-9514, 612-428-2199).

The two right brackets are joined by a rod that adjusts for different-sized templates, and both are tapped for ¼-20 screws to mount the templates. The left bracket is fitted with a free-turning bolt that connects it with the template holders. Locking the left bracket only and turning the adjustment bolt moves the template .05 in. per turn. Recently, I got my hands on some FastTrack aluminum extrusions (available from Garrett Wade Co., 161 Avenue of the Americas, New York, N.Y. 10013; 800-221-2942), which when combined with their micro-adjuster and two micro-blocks made a nearly ready-to-use template holder.

For the template follower, I had to add a means of inserting the bearing into the mortise template, so I devised a keyhole-shaped slot (see the photo on p. 41). After the follower bearing is slid into the narrow slot, it can then be cranked reliably, up or down, into position.

Routing mortises and tenons

Unlike the Leigh dovetail router jig, which uses adjustable templates, my jig has interchangeable templates. I make my templates from scraps of medium-density fiberboard (see the bottom left photo) because it is dimensionally stable, wears well and is easy to work. I make the mortise template first and then shape the tenon template to fit snugly into a test mortise. This is necessary because of the bearing system I use. Instead of ball bearings, I used a bronze bearing that slips on the ¼-in. follower shaft. I matched a ¼-in. mortising bit to a ⁵⁄₁₆-in. bronze bearing. This combination produces mortises that are slightly smaller than the template, so I fit the tenons to the mortise rather than to the template.

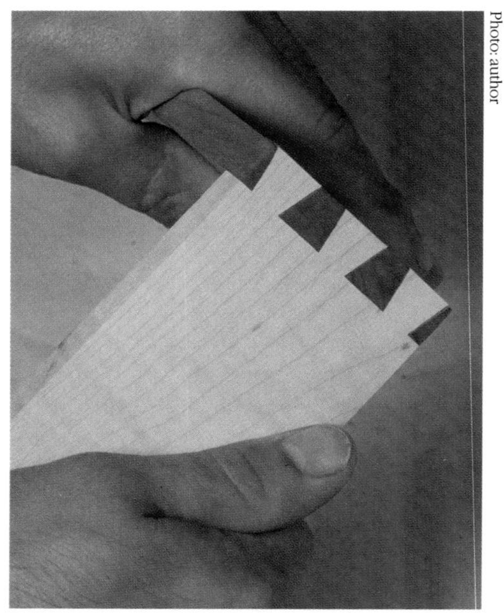
Photo: author

Machine-cut joint speaks for itself— *Perez holds a drawer corner of oiled cherry and maple, dovetailed with his router jig. The jig cuts uniform dovetails or asymmetrical ones, if a hand-cut look is desired.*

I usually eyeball the position of the template by first marking the stock, clamping it in place and then positioning the template holder and follower so that the router bit just grazes my layout lines. Then I'll take a shallow test cut and measure the location with dial calipers. When I cut the stock, the surfaces that will be exposed are face down. I adjust the horizontal position by locking the left template bracket and turning the adjustment screw to move the template. I measure with my dial calipers to compensate for exactly half of the initial error. A similar technique adjusts the follower vertically.

When cutting mortises, the end of the stock bears against the router carrier. The edge is clamped against the fence with the hold-down. An aluminum plate fastened to the upper frame and scribed with a vertical indicator line marks the center of the cut. To lay out my mortises, I mark their center and align them with the indicator.

To cut the tenons, I bolt on a tenon template and change over to a ½-in. straight bit. I climb-cut the first pass, which virtually eliminates any tearout and provides a very clean shoulder line. I complete the cut by merely following the template until no more shavings are produced. The X-Y carriage isn't stiff enough to entirely resist deflection, so I have learned merely to follow the template rather than force the follower bearing against it. I test-fit each piece immediately after machining, and

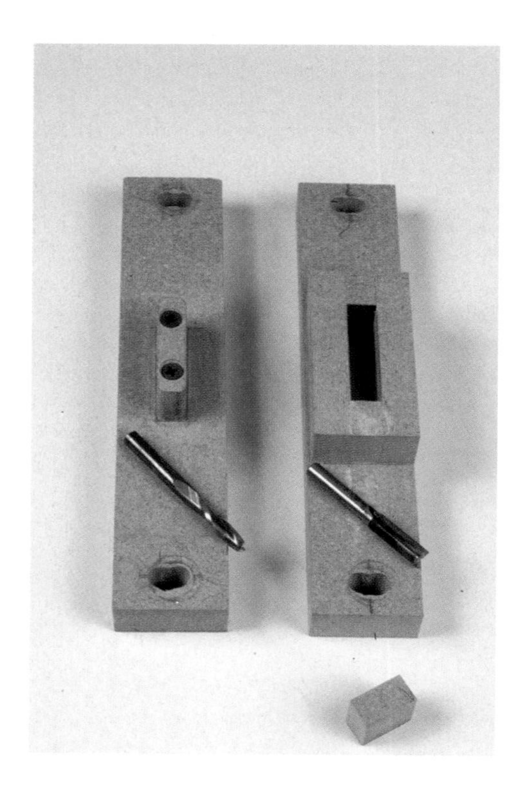

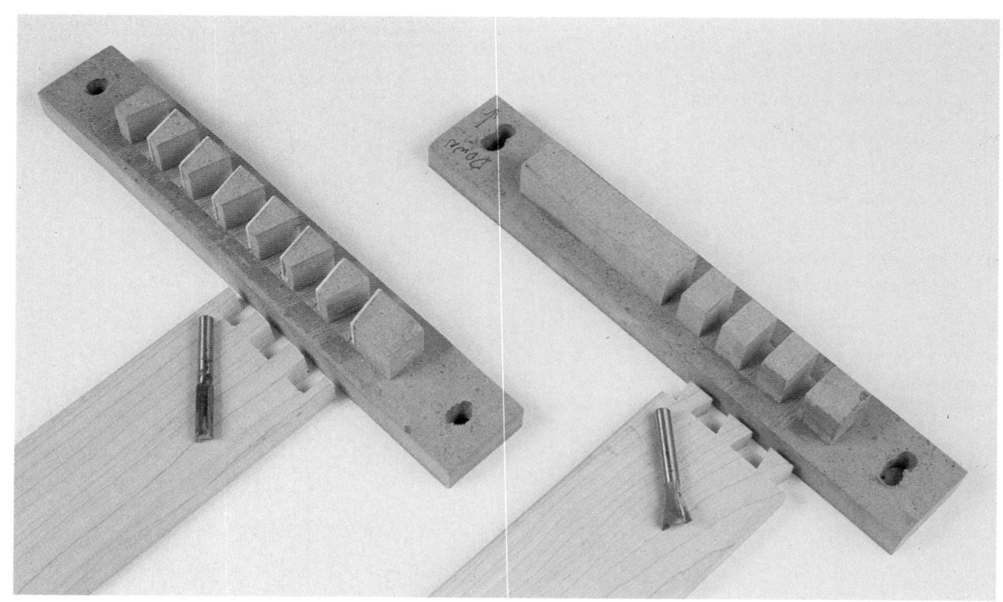

Matching templates and bits— *As a sample of his jig's versatility, the author displays mated templates and corresponding bits. For adjustment, the mortise-and-tenon pattern (left) has a screw-on tenon and a mortise-shortening insert. The pin-and-dovetail templates produced the joinery examples above. When indexed by the tail pattern, the jig can also cut finger joints; simply swap a straight router bit for the dovetail bit.*

I correct a too-tight fit by exerting a little more force during the cut.

Routing dovetails and pins

My joint-making jig handles through-dovetails (see the top photo on the facing page) as easily as it does mortises and tenons. But making a set of dovetail templates is a bit more involved. The main trick is getting the spacing of the pin template to exactly match that of the tails. The tail template is really just a spacing guide, resembling half of a finger joint. In addition to getting the spacing and cut angle right, the pins of the template must be left full enough to ensure a tightly fitting joint. Also, by making the template at least twice as wide as the thickness of the stock, you can adjust the fit of the joint simply by changing the vertical position of the router bit on the stock. Because the height is relative to the template, it's easy to adjust the vertical position of the follower.

I've adapted Mark Duginske's method for cutting dovetail templates on the tablesaw (see *FWW* #96, p. 66). I use a set of wooden blocks to establish the spacing of the dado cuts. After cutting the tail template, I use the same set of blocks to machine the pin template. With this method, you can also make templates for non-uniformly spaced joints as long as you number and order both templates.

Always cut the tails first, using whatever dovetail bit the template is designed for (see the bottom right photo on the facing page). I place a piece of stock flat against the router carrier to set the depth of cut, extending the bit just proud of the piece, and clamp the stock face down. I adjust the template holders horizontally and position the cut so the outside tails are equidistant from the edges of the board.

I fit my router with a straight bit for machining the pins. I mount the template and position the stock so that the inside of the joint faces down. This arrangement ensures that once everything is adjusted, slight variations in stock thickness will not affect the joint's fit. The fit is determined by the distance between the follower and the bit—smaller distances will yield tighter joints and vice versa. With a test piece clamped in place (and the router unplugged), I position the follower so that the bit is just below the workpiece when the follower first contacts the bottom of the template. From here, I make test cuts and raise the follower. □

Guy Perez is studying political philosophy in Madison, Wis. He also builds furniture for his family and friends.

A shopmade hold-down

I originally used Jorgensen adjustable bar clamps to hold down workpieces on my joint-making jig. But I soon found the repeated tightening and loosening of the clamps to be time-consuming and a real blister maker. I also dismissed the idea of using toggle clamps because of their small size. Instead, I constructed my own clamp using scraps of hard maple, a piece of medium-density fiberboard (MDF) for a base, ⅝-in. threaded rod, dowels, screws and an assortment of 5/16-in. bolts (see the photos on p. 39).

Building a hold-down is straightforward once you understand the basic operating principle. In the vertical clamp shown in the drawing below, the handle provides leverage to the clamping arm by means of a pivoting bracket, which is fixed between the arm and handle. The clamp locks in place when the handle's pivot point is pulled forward of the arm's pivot. But because clamping pressure diminishes as the arm pivot travels past the initial locking point, a travel-limiting stop is needed. The trick is in placing the stop so that the clamp locks down and exerts sufficient pressure. I arrived at a good balance (favoring clamping strength) by positioning and paring the block (crossbar) until I was satisfied with the locking action.

Because the forces in the hold-down are mostly vertical, I oriented the grain of the base bracket up and down to prevent splitting. However, because the bracket is screwed to the base, it's possible that the drywall screw could pull out of the bracket's end grain. To counteract this, I reinforced the base brackets with hardwood dowels. Holes drilled through the base enable me to bolt the hold-down to my jig's table. A pair of adjustable spindles with clamping pads resist any side-to-side movement of the workpiece. The spindles are two lengths of ⅝-in. threaded rod with top and bottom nuts. Rubber cap protectors (available at most hardware stores) serve as the pads.

I use my oversized toggle clamp almost exclusively as a hold-down for my router jig, but I sometimes use it as a helping hand when I am power-sanding or freehand-routing. The clamp exerts a lot of down pressure, and I can quickly reposition the stock. The greatest virtue of the clamp, however, is its sheer size. Its long reach and big handle make the clamp truly a pleasure to use in repetitive operations. *—G.P.*

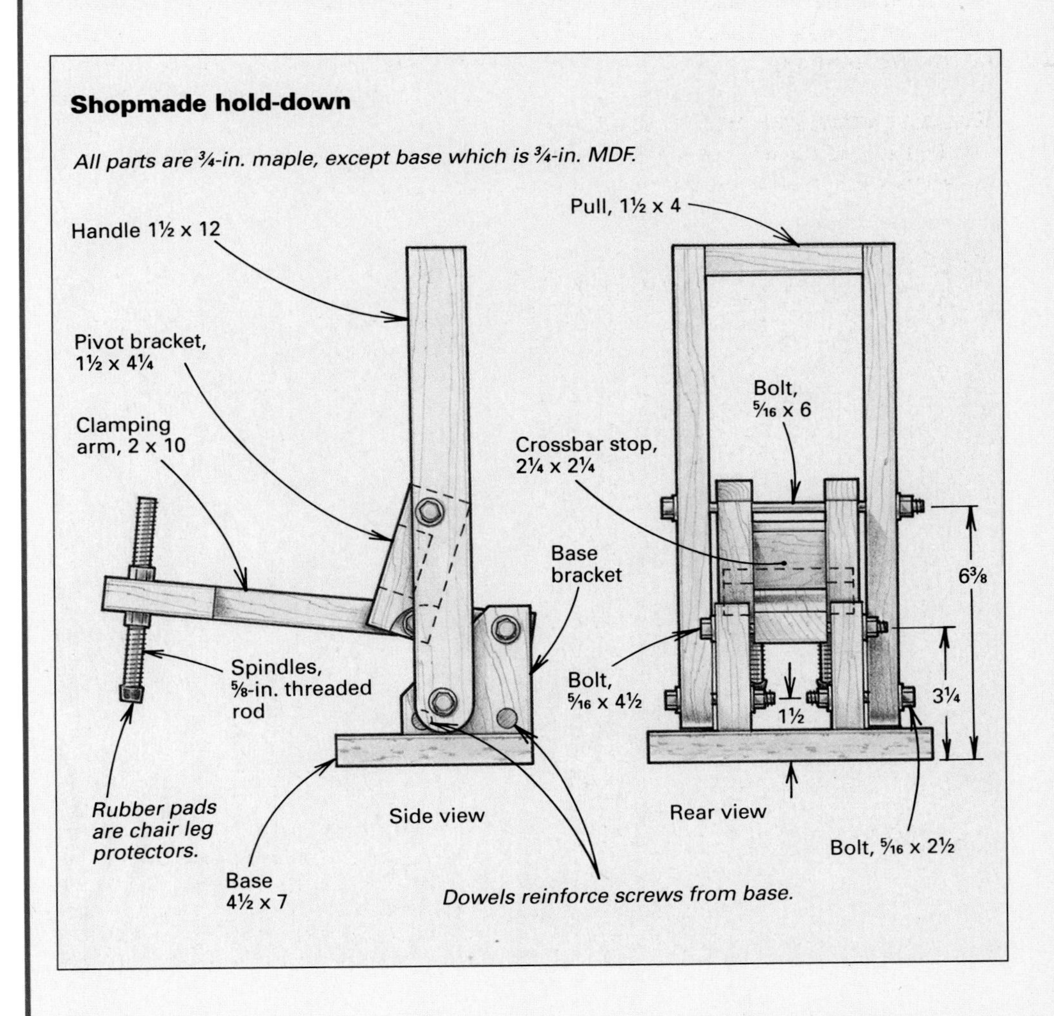

Vacuum Hold-Down Table

Knockdown sanding and routing platform grips work and controls dust

by Mike M. McCallum

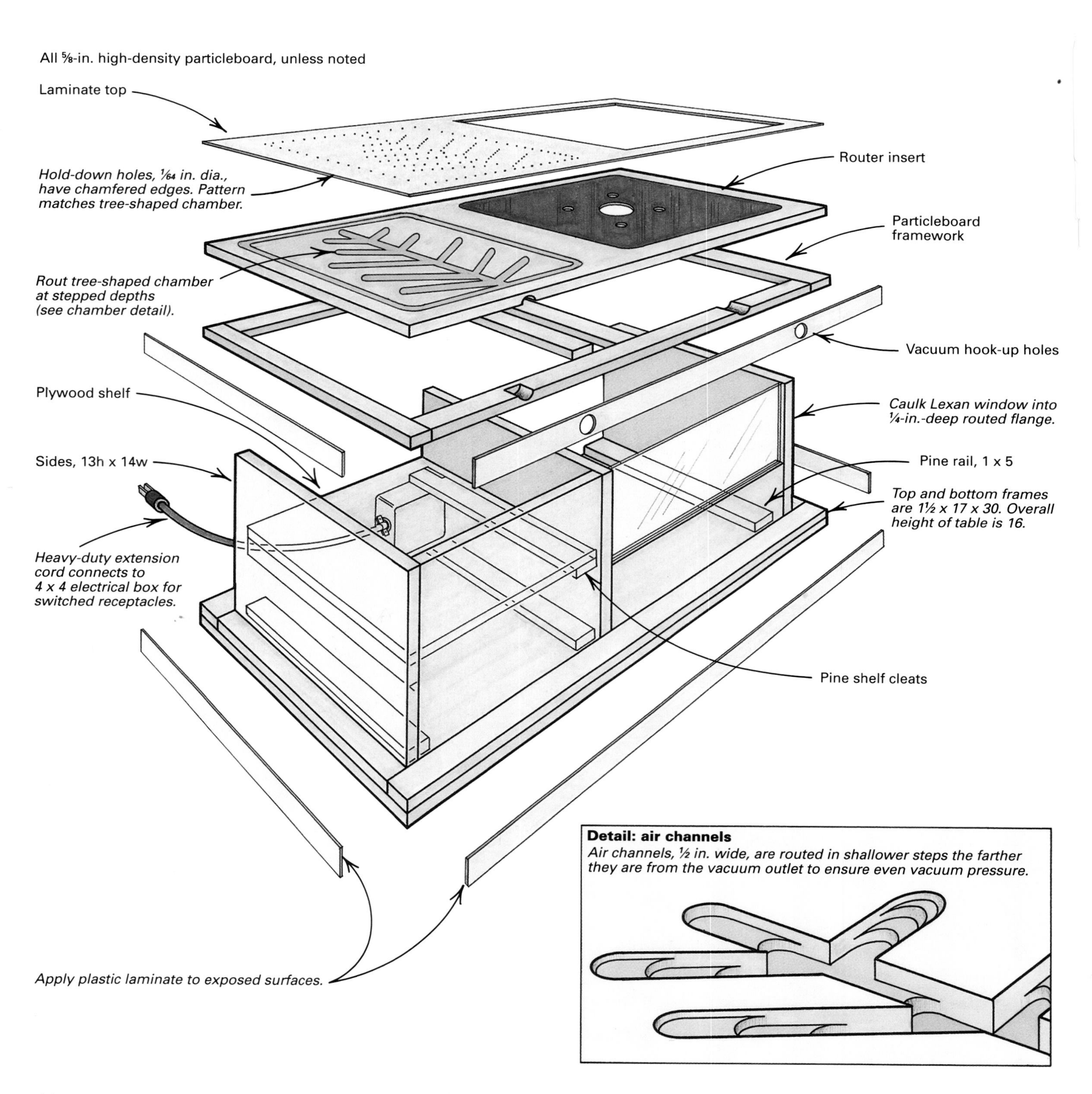

From *Fine Woodworking* (March 1993) 102:68-69

When I'm constructing a set of custom cabinets, I frequently need an extra pair of hands, especially when I'm sanding drawer fronts or drilling odd-shaped pieces. Occasionally, I also need a table-mounted router. More often than not, I require that router table or those pair of hands at a job site. After putting up with cobbled scraps, makeshift clamps and excessive router dust one too many times, I came up with a design for a router table that's also a vacuum hold-down. Using scrap materials, I built the table so that I could easily disassemble it for storage or transport.

I call my knockdown platform a super router/hold-down table for a couple of reasons. First, it's stout, turning my router into a light-capacity shaper. Second, it enables my shop vacuum to serve dual functions by providing suction for the hold-down surface or collecting dust from the router table. And while I don't use the hold-down to freehand-rout large workpieces, I do rely on its substantial holding power for most of my sanding and finishing work (see the top photo).

Vacuum side has plenty of clamping power*—The vacuum surface (left) clamps a drawer front as the author sands its face. An open hole on the edge of the table (right) shows where he connects his shop vacuum when he's routing.*

To wire his table, *the author fed a 4-in. work box for a pair of switched receptacles: one for the vacuum, one for a router or sander. A shelf at the back of the table holds tools and accessories. He drew outlines of the items and routed recesses to hold each shape, which reminds him when something's missing.*

Design and construction—The dimensions of the hold-down table are not critical, but be sure to adjust for the size of your work area, vacuum hose and router. I made my table out of ⅝-in. high-density particleboard and covered exposed surfaces with scraps of plastic laminate. The top is removable, so I can use the vacuum table on my benchtop. I stiffened the table's top and bottom by gluing on a particleboard framework, as shown in the drawing on the facing page. The top and bottom frames hold the sides and center divider in place without fasteners, allowing easy knockdown of the unit. After assembling the top and bottom oversized, I trimmed the parts square. I laminated all the pieces, and then I bored two holes in the edge of the table, so I can connect my shop vacuum to either the router-table or hold-down side (see the top photo). To power the table, I ran a heavy-duty extension cord to a 4x4 electrical box and mounted the box's stud bracket to the inside of the platform. The box houses switched receptacles for both the router (or sander) and shop vacuum. I also added a plywood shelf (see the bottom photo) to the table to hold tools, bits, guide bushings and adapters. I ordered most of these accessories through MLCS Ltd. (P.O. Box 4053 C-13, Rydal, Penn. 19046).

The router-table side—I made a clear window for the router table from Lexan, which I recycled from a computer-store display. The window is a good safety feature because it lets in enough light to see the collet when I'm adjusting the bit height or using the router table. The router insert is a standard one—it fits whatever bit I'm using. Oak Park Enterprises, Ltd. (Box 13, Station A, Winipeg, Manitoba, Canada R3K-129) carries complete bearing and insert kits for various router models. To prevent vibration and flush the router-table surface, I oiled the insert's bearing surface and then caulked the face of the insert with silicone before I placed it into the flange that I had routed into the tabletop. I clamped the assembly flat on my tablesaw. In use, I estimate that my 5-gal. collector sucks up about 50% of the dust from the router. I'm sure that with a few provisions, such as adding intake holes, I could improve the dust-collection capability considerably.

The hold-down side—On a piece of tracing paper, I drew an evenly spaced tree-shaped hole pattern that was suitable for my hold-down needs. After I transferred the tree pattern onto the particleboard top, I freehand routed the air channels for the vacuum chamber. To ensure an even vacuum across the hold-down surface, I routed the channels at ascending depths (see the air-channel detail in the drawing). I based the stepped depths on vacuum-drop ratios for the chamber volume. If you're using plywood for the table, paint or seal the routed pattern to prevent air leaks before you glue on the laminate. To make the hole pattern in the laminate, I first placed a clear piece of plastic over the routed chamber and poked out a hole pattern to follow the air-channel shape. Then I used a crayon to rub the hole locations onto the laminate. When boring through the laminate, use a tiny bit (I used a ¹⁄₆₄-in. twist drill). The small orifices, through the Venturi principle, increase the vacuum. Finally, lightly countersink the holes.

Using the vacuum hold-down—As long as my workpiece has a flat surface to put down on the hold-down table, I've found that there's plenty of suction—enough to grip a piece of low-grade plywood. To increase the holding pressure, you could also block off holes that are not covered by the workpiece. On rough surfaces, I take a ⅛-in.-thick piece of closed-cell plastic (shipper's foam) to make a gasket. With a utility knife, I cut out an appropriate shape that still allows the vacuum to suck the workpiece down. I use a couple of pieces of masking tape to hold the gasket to the table. If you need to hold down a sphere, an odd shape or a piece with a very uneven surface, you can make a holder as follows: Sculpt out a Styrofoam gasket for the shape you want to secure. In a well-ventilated area or outdoors, heat up a piece of nichrome or small stainless-steel wire with a propane torch, so you can make a series of holes through the gasket. Tape the Styrofoam to the hold-down table, and you've got a fairly quick clamp to hold just about any shape that you have to sand or drill holes in. So far, I've been delighted with the possibilities of the hold-down table. In fact, I'm working on a sliding saw table that uses a similar vacuum hold-down system. □

Mike McCallum is an artist who does custom architectural woodworking in Portland, Ore.

Photos: author; drawing: Matthew Wells

Vacuum Motor Turns into a Spray Rig

Enjoy the benefits of high-volume, low-pressure finishing in a compact unit

by Nick Yinger

For years, I did my spray finishing with a conventional compressor-driven setup. I was never entirely satisfied with the arrangement, and I recently built my own high-volume, low-pressure (HVLP) unit, as shown in the photo at right, to replace it. What bugs me about conventional spraying? For starters: finishing the inside of a case with a swirling cloud of overspray billowing back in my face. I can't see what I'm doing, and I wind up ingesting a big dose of chemicals no matter what kind of mask I wear. Even when I'm spraying water-based finishes, which are inherently safer, I find overspray annoying. Although they're neither toxic nor flammable, water-based finishes are expensive, so it makes even less sense to blast these precious fluids all over the booth with air compressed to 50 pounds per sq. in. (psi). HVLP spraying looked like the answer to these problems. This method promised to transfer 70% to 80% of the material from the gun to the object compared with 20% to 30% with a conventional setup. To accommodate a stream of warm, dry, low-velocity air, HVLP guns have large hoses and air passages. They use copious amounts of air—as much as 30 cu. ft. per minute (cfm) but at only 5 psi. (For pros and cons of HVLP, see the box on the facing page.)

I had a 3-hp compressor, so it seemed a simple matter to install a large, low-pressure regulator to feed 5-psi air to the gun. But there was a catch. A 3-hp piston compressor won't pump 30 cfm continuously at any pressure. The rule of thumb is 1 hp per 4 cfm of air, and we're talking about large, healthy, industrial horses not puny, underfed, home-improvement horses. Because 8- to 10-hp compressors are expensive and connecting my small compressor to a tank the size of a submarine seemed impractical, I decided I'd investigate the turbine compressors sold with HVLP guns.

I borrowed an HVLP unit from a friend and used it to finish some bathroom cabinets. It performed beautifully: almost no overspray, good atomization and good fluid and pattern control. My only criticisms were that the hose seemed cumbersome, and the handle of the gun became uncomfortably hot.

As I used the HVLP unit, I couldn't help thinking that if it acts like a vacuum cleaner, sounds like a vacuum cleaner, it must *be* a vacuum cleaner. I peeked inside. Sure enough—a two-stage vacuum cleaner turbine with an 8-amp motor! Soon thereafter, I set out to build my own HVLP turbine compressor.

Build your own HVLP unit

An HVLP machine is a centrifugal turbine compressor contained in a box with an inlet to bring air into the turbine and a plenum or outlet chamber to capture the compressed air discharged by the turbine and route it to your sprayer hose (see the drawing on p. 49). The turbines used in large vacuum cleaners are integral with their electric motors and are referred to as vacuum motors.

Shop-built spray unit—A high-volume, low-pressure unit like this one that the author built is ideal for on-site work or in the shop.

First buy a vacuum motor—Go to an industrial supply company, or get their catalog. I bought mine at Grainger (contact their marketing department at 333 Nightsbridge Parkway, Lincolnshire, Ill. 60069; 800-473-3473 for the nearest location); their catalog lists 45 vacuum motors, ranging from $40 to more than $280. You'll find a wide selection of features, such as bearing type, motor voltage, number of compressor stages and motor amperage. Most important for this application is *bypass*, not flow-through motor cooling. This means the motor is cooled by a separate fan. With this design, the motor won't overheat if the vacuum inlet or outlet is obstructed.

Single-stage compressors move large volumes of air but produce the lowest pressure. Two- and three-stage units supply higher

Photos: Jonathan Binzen

Conventional spraying vs. HVLP

by Dave Hughes

Okay...it's 8 a.m., and you've just entered your shop, coffee in hand. Standing before you is your latest project, nearly completed. It just needs to be lacquered. You take a deep breath, fill your spray gun, crank up the compressor, put on a particle mask and go for it. Fifteen minutes later, the atmosphere in your shop resembles that of Venus, every tool is covered with a fine white dust, the shop's out of commission for the rest of the morning and you've got a serious headache. Sound familiar? If, like most of us, you've tried to do finishing with conventional spray equipment in a small shop space, it probably does. Well, there's an alternative. It's high-volume, low-pressure (HVLP).

By now, most professional finishers have an HVLP unit in their arsenal of tools and increasingly, the units are finding favor with folks who do only occasional finishing. One big reason is that HVLP units have far higher transfer efficiency than conventional spray units. This means, simply, that most of the stuff you're spraying goes where you want it to go. A painter friend of mine did his own little test when HVLP first hit the market. He painted one cabinet with a traditional, compressor-driven gun and an identical cabinet with an HVLP unit. When he was done, there was three times as much paint left in the HVLP cup. Where was the paint missing from the conventional gun? All over.

Aside from transfer efficiency, HVLP offers a string of clear benefits over conventional setups:

- They are compact, lightweight, self-contained, easy to set up and clean.
- The guns have a wide variety of spray-pattern settings for finishing intricate shapes as well as broad, flat surfaces.
- The low-pressure air supply is adjustable and so creates far less bounce-back of material from inside corners.
- The dry, heated air helps materials flow on smoothly, level out nicely and set up quickly. It also helps avoid blushing on cold, damp days.
- Your shop is not rendered useless for hours. (But open a window anyway.)

Drawbacks? There are a few:

- HVLP units are not really a high-production tool but are more suited for small- to medium-sized projects.
- Standard models have a rather cumbersome air hose all the way to the gun, limiting wrist mobility somewhat.
- As with any quart-gun arrangement, you can't spray upside down, and you're constantly, it seems, filling it up. (Higher priced models offer a 1- or 2-gal. pot that stands on the floor for less-restricted gun movement and less-frequent fill ups.)
- And there's that whining motor—it reminds me of a car wash vacuum.

HVLP is a definite advance for the small-shop woodworker or finisher who wants professional results. With prices starting under $500 and savings from high transfer efficiency, they're a good investment. From the money you save, stake yourself fifty bucks for a decent charcoal respirator and a pair of earplugs. □

Dave Hughes is a professional finisher in Los Osos, Calif.

High volume, low pressure (HVLP) in a small package. *At 15 in. sq. and 18 lbs., the shop-built turbine-powered HVLP spray unit in the photo below is a fraction of the size and weight of the standard medium-sized compressed air setup in the photo at left.*

pressure air at some sacrifice in volume but typically have more powerful motors and, hence, better overall performance. I chose a two-stage turbine with a 13-amp motor rated at 116 cfm that costs $163, an Ametek model #115962. I could have purchased a less powerful unit, but I wanted to be able to operate two spray guns on occasion, and anyway, I like overbuilt machinery. For a one-gun setup, you might try the Ametek 115757-P, which costs $63. For the rest of the parts in my HVLP unit, including the hose but not the gun, I spent less than $70.

Make a cradle for the motor—These motors are designed to be mounted by clamping the turbine housing between two bulkheads using foam gaskets. Make the rear bulkhead first. Cut it to size, bandsaw the circular hole and then chamfer the back side of the hole. The chamfer will ease the flow of motor-cooling air away from the motor housing. Cut the positioning ring to size, and rough out the hole with the jigsaw, leaving it slightly undersized. I made a Masonite routing template to exact size by cutting the hole with a fly cutter on the drill press. Use the routing template to finish the hole in the positioning ring.

Cut the housing sides, top and bottom to size, and make the dado for the rear bulkhead in each of them. Then drill the cooling outlet holes in the side pieces. Assemble the housing with the rear bulkhead in place, and when the glue has set, drop in the posi-

Improving sprayer output—*Plastic laminate coiled in the outlet chamber acts as a fairing and increases output by lowering resistance. Weather stripping and rubber tubing form gasket seals.*

Mounting electricals—*Switch, cord and circuit breaker are mounted in the back panel. Holes in the side of the back chamber are for motor-cooling air. A wooden cleat holds the wound cord.*

tioning ring, and glue it in place. I used screwed butt joints for the housing pieces and relied on the bulkhead to stiffen the box.

Gasket and sealant—The turbine is held in the circular rabbet created by the bulkhead and positioning ring and is isolated from the wood by silicone rubber sealant. To hold the turbine centered in the rabbet while the silicone sets, cut three 2-in.-long pieces of ⅛-in.-inside-dia. (ID) soft rubber tubing that compresses to about 1/16 in. under moderate pressure. (This surgical tubing, with a wall thickness of 1/32 in., is available in hobby shops and medical supply houses.) Lay the housing on its back, and put a generous bead of silicone in the rabbet. Lay the three pieces of tubing across the rabbet at 12 o'clock, 4 o'clock and 8 o'clock, and push the turbine down into the wet silicone. If you want the turbine to be easily removable later, spray the rim with an anti-stick cooking spray such as PAM before setting it into the silicone. Let the silicone set, and trim off the squeeze-out and tubing ends later.

With HVLP, far more of what you spray sticks to the object you're spraying.

Next rout the gasket grooves around the front edge of the housing, and press lengths of 3/16-in.-ID soft rubber tubing into them. Make the front and back covers, and apply the rings of ½-in.- by ½-in. adhesive-backed weatherstrip, as shown in the photo at left, and then screw on the front and back.

Holes in the box—I tried various locations for the outlet holes and found no detectable differences. But I did get better output when I installed a fairing made from a strip of plastic laminate, which makes the outlet chamber roughly cylindrical (see the photo at left). Drill one or two 1-in. outlet holes in the housing, and screw ¾-in. pipe thread close nipples into them. Attach adapters to the nipples to provide ¾-in. male hose threads.

I attached a large shop-vacuum air filter to the front cover. Four short dowels hold the base of the filter in place, and a bracket pulls it tight against the cover. The bracket consists of two threaded rods screwed into the front cover joined by a hardwood crosspiece with a bolt through its center. A washer and wing nut secure the closed end of the filter against the crosspiece. You could also try using a large automotive filter. In that case, a Masonite or plywood disc secured by a similar bracket could hold the filter against the front cover.

Electricals—Mount the electrical parts: a heavy-duty switch, a circuit breaker with the appropriate rating for your motor, and the supply cord through the back cover, as shown in the photo at right. Then add rubber feet, a carrying handle and a cord-storage device.

Nice hose—I tried three different types of hose. All were ¾ in. ID and can be equipped with ordinary garden hose threaded fittings or quick-connect couplers. The most flexible was the lightweight, corrugated type provided with most factory-built HVLP sprayers, but its rough inner surface doesn't deliver as much air as smoother types. Plastic garden hose is cheap, smooth inside and flexible when warm, but in use, the heated air causes the hose to become too soft and to kink easily. My favorite is Shields Vac extra heavy duty/FDA hose available from marine distributors. It is made of a soft flexible vinyl molded around a hard vinyl helix. It's recommended by the manufacturer for use in boat plumbing below the water line, which means it will withstand a lot of heat as well as mechanical and chemical abuse.

Gun control—You can't just hook up your old gun to your HVLP turbine. HVLP guns are designed to enable them to atomize fluids with low-pressure air. List prices for these guns start at around $250. Of the HVLP guns I've tried, my favorite is a DeVilbiss (contact DeVilbiss at 1724 Indian Wood Circle, Suite F, Maumee, Ohio 43537; 800-338-4448 for a local supplier). The current model most like mine is their JGHV 5285 that lists for $365. It has stainless-steel fluid passages and a stainless-steel needle, so water-based finishes won't cause corrosion. And much to the relief of my palms, the handle is a nylon composite that doesn't get hot in use. □

Nick Yinger is a professional land surveyor in Kirkland, Wash.

Shopmade HVLP unit

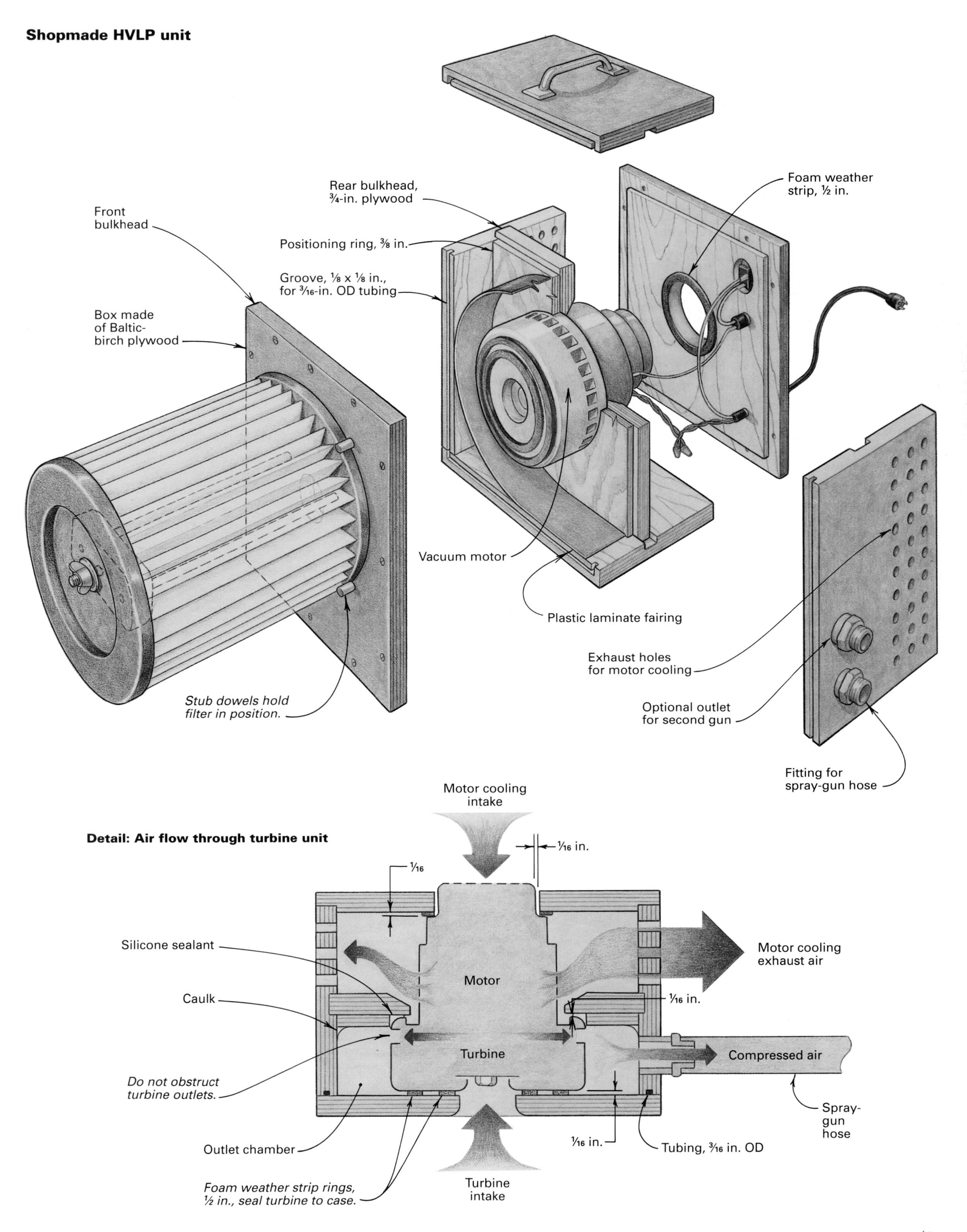

Drawing: Heather Lambert

Vacuum Powered Hold-Down

Look ma, no clamps

by Evan Kern

Evan Kern built a vacuum table to hold down thin stock *when he's rotary planing with his drill press. To increase the table's suction, he covers holes ahead of the cutter with cardboard. Here, Kern advances the cardboard with a walnut workpiece as he guides it along a fence that's clamped to the table.*

As an avocational instrumentmaker, one of my challenges is planing resawn wood to less than ¼-in. thick. My attempts to thickness wood with a conventional planer usually result in hopelessly warped or shattered pieces of wood. Although an abrasive planer can do the job, one of those machines is well beyond my financial means. And since my needs for thin stock are modest, I bought a Wagner Safe-T-Planer, which is an inexpensive rotary planer that I chuck in my drill press.

The only drawback I encountered while rotary planing stock was the tendency for the wood to lift up, especially at the beginning and end of a pass, resulting in pieces that were unevenly thicknessed. To solve this problem, I built a vacuum hold-down table for my drill press, as shown in the photo at left. The hold-down surface's holes go through the tabletop and into a labyrinth (vacuum chamber), which is connected to an ordinary shop vacuum. The vacuum holds thin stock flat against the table, enabling me to plane pieces down to ⅓₂ in. and up to a ½-in. maximum thickness. Although I use my hold-down table for planing, I suspect that with a few modifications to clamp it to a benchtop, the table could be used for light routing and sanding.

Constructing the hold-down

The hold-down table consists of a ¾-in.-thick medium-density fiberboard (MDF) tabletop mounted to a hollow base. The drawing shows the size and pattern of the holes to bore through the top. Three pieces of ¾-in. plywood—the center one being the labyrinth—make up the base. A ½-in. plywood bottom is screwed to the base to provide ears for clamping. Two requirements that may be different for other drill-press tables and shop vacuums are the size of the bottom (mine is 15 in.) and the size of the vacuum opening (mine fits a 1⅛-in.-OD PVC coupling).

Labyrinth—In addition to joining the holes in the table to the vacuum source, the labyrinth supports the workpiece beneath the cutter. The suction from even a small vacuum can distort the table if it's not adequately supported. After scroll sawing out a labyrinth (see the pattern in the drawing), cut out the other two base pieces and sandwich and glue the labyrinth between them. After the glue has dried, drill a hole for the vacuum hose.

Adding a control gate and a fence—If all the holes in the hold-down table are covered by a workpiece, there will be no relief for

Photos: Alec Waters; drawing: Mark Sant'Angelo

the vacuum and, as a result, your vacuum's motor may overheat. You can eliminate this problem by making a vacuum-control gate, which allows air to enter the labyrinth. I made a simple gate (see the photo at right) out of ¼-in. plywood. The gate slides over a pair of ½-in. holes bored in one side of the base. I can open the gate fully or partially to equalize the pressure in the labyrinth and regulate the degree of suction at the hold-down surface.

To guide stock when planing, I made a plywood fence that I spring clamp to the tabletop. The fence has a recess that lets the Safe-T-Planer overlap the edge of the work. I faced the underside of the fence with 1/32-in. plywood to cover the holes that would otherwise be exposed by the recess and to provide an edge for the workpiece to ride against at the recessed area.

Before mounting his Safe-T-Planer, *the author chucks in a bent piece of wire to level the hold-down table. While hand-turning the chuck, he observes and shims the table until the wire's tip grazes the surface for a full revolution. After connecting the vacuum's hose to the front of the table's base, he opens the vacuum-control gate on the base's side. Then, using discs of modelmaker's plywood to gauge thickness, Kern will set the planer's depth of cut.*

Rotary planing on the hold-down table

A Safe-T-Planer consists of a shaft connected to a 3-in.-dia. disc that holds three circular cutters (see the photo at right). Because the cutters only project about 1/64 in. from the disc, these rotary planers are quite safe. The planers, which will also work in most radial-arm saws, are manufactured by G & W Tool, Inc., P.O. Box 691464, Tulsa, Okla. 74169; (918) 486-2761 and are available at most woodworking supply stores. When using a rotary planer, the length of stock that can be planed is limited only by your shop space. The stock width is limited to your drill-press swing.

Squaring the table and setting the cut—To make sure the hold-down table is perpendicular to the drill-press spindle, I made a gauge by bending a heavy piece of wire into a Z-shape. After I mount the wire in the chuck, I rotate the chuck by hand and observe the gauge and the table's top. The gauge's tip should just touch the table's surface throughout its rotation (see the photo above).

I use 3-in.-dia. plywood discs as thickness gauges to set the height of the planer's cutter above my hold-down table. I bandsaw the discs from sheets of modelmaker's plywood (available at most hobby shops), which comes in precise thicknesses from 1/64 in. to ½ in., in 1/64-in. increments. After placing a disc of the desired (planed) thickness on the table, I adjust the quill until the cutter just touches the gauge, and then I lock the quill.

Feed and cutter speed—After lightly waxing the tabletop, I hook up my shop-vacuum hose and turn my drill press on. If the wood is wider than the planer, after an initial pass, I reverse it end for end and continue passes, moving the fence in toward the drill-press column until I've planed the entire width of the board. I feed stock at a rate of approximately two to three square feet of wood per minute. At this rate, I've never had to sharpen the cutters, even though the manufacturer supplies instructions for this. Although the planer's maker recommends speeds of 3,000 RPM to 6,000 RPM, I've found that 2,300 RPM helps prevent the cutters from burning the wood during those inevitable feed pauses.

Regulating the suction—When I'm feeding narrow strips of wood into the planer, most of the hold-down table's holes are uncovered, and as a consequence, there's an insufficient vacuum. I resolve this by covering exposed holes with pieces of cardboard or stiff plastic. Feeding work against the planer cutters (from left to right) pushes the covers out of the way (see the photo on the facing page). As the end of a board is reached, I reintroduce another cover so that the holes in the table are continuously covered to maintain a vacuum. □

Evan Kern is an author and a retired dean of Kutztown University. He builds stringed instruments and puzzles in Kutztown, Pa.

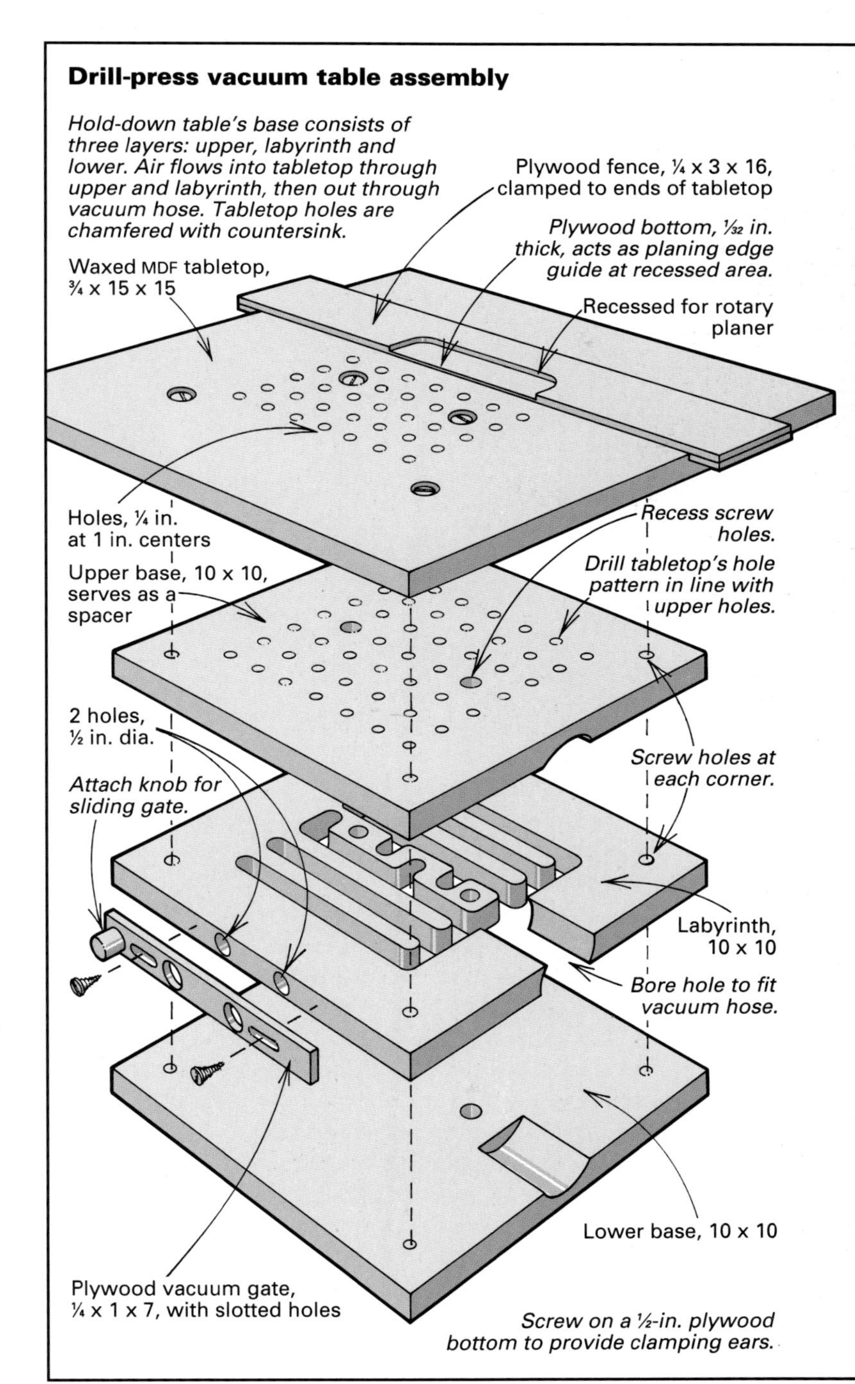

From *Fine Woodworking* (November 1992) 97:82-83

One-Stop Cutting Station

Tablesaw jig handles crosscuts, tenons and miters

by Ken Picou

***Making a crosscut box more versatile—** An accurate sliding-crosscut box makes a good base for cutting accessories, including this corner-slotting jig. This jig mounts or dismounts in seconds and makes for strong miter joints in picture or mirror frames and in small boxes or drawers.*

Tablesaws are excellent for ripping stock, but the standard miter gauge that comes with most tablesaws makes them mediocre at best for crosscutting material or cutting joinery. But by making a simple sliding-crosscut box and a few accessory jigs, you can greatly increase the accuracy and flexibility of your saw and turn it into a one-stop cutting station, capable of crosscutting, tenoning and slotting.

The system I've developed consists of a basic sliding-crosscut box with a 90° back rail, a removable pivoting fence, a tenoning attachment and a corner slotting jig, for cutting the slots for keyed miter joints (see the photo above). This system is inherently safer and more accurate than even the most expensive miter gauge for several reasons. First, it uses both miter slots, so there is less side play than with a miter gauge. Second, the work slides on a moving base, so there's no chance of the work slipping or catching from friction with the saw table. Third, the long back fence provides better support than a miter gauge, which is usually only 4 or 5 in. across. Fourth, the sliding-crosscut box is big, so angles can be measured and divided much more accurately than with a miter gauge (the farther from its point of origin an angle is measured, the greater the precision). Finally, the sliding crosscut box is a stable base on which to mount various attachments, such as a tenoning jig or a corner slotting jig, which can greatly expand the versatility of the tablesaw.

Building the basic crosscut box

I cut the base of my sliding-crosscut box from a nice, flat sheet of ½-in.-thick Baltic-birch plywood, and then I make it a little bit wider and deeper than my saw's tabletop. A cheaper grade of plywood also would be fine for this jig, but I decided to use a premium material because I wanted the jig to be a permanent addition to my shop.

The runners that slide in the tablesaw's miter-gauge slots can be made from any stable material that wears well. I prefer wood to

Photos: Vincent Laurence

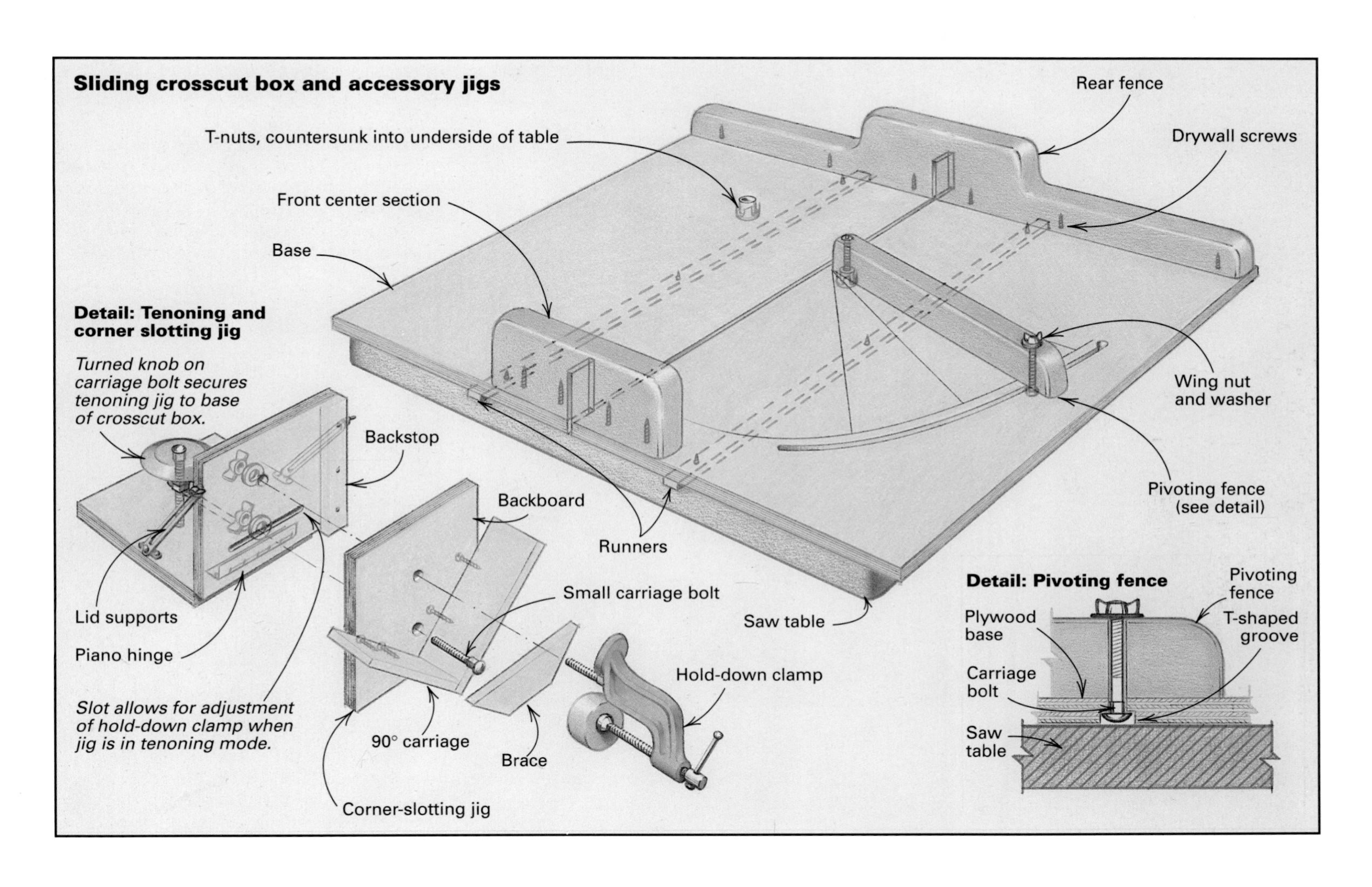

metal because wood works easily, and I can screw right into it. I usually use hard maple, and I've never had a problem. Using a long-wearing, slippery plastic such as an acetal (Delrin, for example) or ultra-high molecular-weight (UHMW) plastic is also a possibility. (For more on using plastics for jigs and fixtures, see *Fine Woodworking* #105, pp. 58-61.)

I start with a maple board of sufficient length that is at least as wide as three or four runners are thick. I plane this board, taking off minute increments with each pass, until it slides easily on edge in one of the slots but isn't sloppy. Once the fit's right, I rip the runners from this board, setting the fence on my tablesaw to just under the depth of the miter-gauge slot. Then I drill and countersink them at the middle and near both ends (I check the dimensions of the Baltic-birch base to make sure I drill the screw holes so they'll fall near the edges of the base). I usually drill a couple of holes near each end as insurance in case a screw drifts off when I'm screwing the runners to the base.

Next I crank the sawblade all the way down below the table and lay the runners in the miter-gauge slots. I position the base so that its back edge is parallel to the rear of the saw table and the front edge overhangs by a couple of inches. I clamp the runners to the base in the front. I drill pilot holes in the plywood from below using a Vix bit (a self-centering drill bit available through most large tool catalogs) placed in one of the countersunk holes in the runners. Then I screw up through the runners into the base. When I've done both runners at the front of the saw, I slide the base back carefully and repeat at the rear (see the top photo on p. 54). I check for binding or wobble by sliding the base back and forth a few times. If the fit is less than ideal, I still have four more chances (the extra screw holes I drilled at both ends of each runner) to get it right. If the fit is good, I drill pilot holes with the Vix bit and screw the runner to the base in the middle, taking care not to let the runner move side to side. I also trim the runners flush with the front and back of the crosscut box.

If the fit's a bit too snug at first, use will tend to burnish the runners so that they will glide more easily. If, after some use they're still a little snug, you can sand the runners just a bit and give them a coat of paste wax. That will usually get them gliding nicely.

Drawing: Michael Gellatly

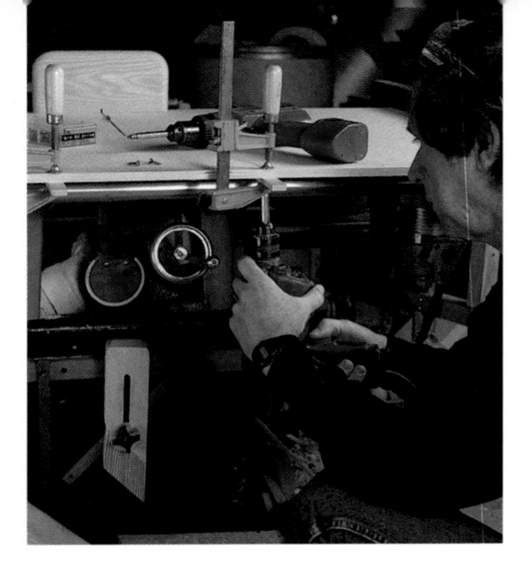

Accurate holes are key to an accurate jig. *Clamps hold the crosscut-box runners in place (left) while the author drills and screws the runners to the base. Using a Vix (self-centering) bit in the previously drilled and countersunk holes in the runner keeps the bit centered going into the plywood, which helps keep the screws from pulling the runners out of line.*

Checking and rechecking for a perfect 90° *(below), both with a square and with test-cuts, is time well-spent. The accuracy of the whole crosscut box and all jigs that mount to it depends on getting the relationship of rear fence to blade just right.*

Rear fence helps align jig's hinge—*Using the rear fence as his reference, the author aligns the tenoning jig's hinge with a square (below). The Vix bit ensures that the screw holes are centered, so the screws will go in true and the hinge will be straight.*

Building accuracy into the jig—An inaccurate jig is useless, so it's essential that assembly of this jig be dead-on. Fortunately, this isn't difficult; it just takes a little time and patience.

I made both the back fence and the front center section from straight-grained red oak, but any straight-grained hardwood will do (see the drawing on p. 53). I make sure the center portions of both pieces are built up high enough to provide 1½-in. clearance with the blade cranked up all the way.

The front section helps keep the table flat and prevents it from being sawn in half. Because this front section is not a reference surface, its position isn't critical, so I screw it on first.

Then I mount the rear fence about ¼ in. in from and parallel to the back of the Baltic-birch base. I clamp the fence to the base and drive one screw through the base, which I've already drilled and countersunk, into the fence a couple of inches to the right of where the blade will run. This provides a pivot point, making it easier to align the rear fence to the blade.

I remove the clamp, raise the blade up through the base and cut through the front section and the base, staying just shy of the rear fence. So far, there's only one screw holding the rear fence in place. To set the rear fence permanently and accurately at 90° to the blade, I place the long leg of a framing square against the freshly made kerf (saw is *off*) and the short leg against the fence. With the fence flush against the square, I clamp the fence on an overhanging edge and do a test-cut on a wide piece of scrap. I check this for square with a combination square and adjust the position of the fence as necessary. When I've got it right, I put another clamp on the fence near the blade on the side opposite my one screw. Then I drill, countersink and screw through the base into the fence right next to the clamp, and I check the fence's position again to make sure screwing it to the base didn't pull it off the mark (see the bottom left photo). I also make another test-cut, and as long as it's still good, I screw the fence down near the ends and the middles on both sides of the blade (see the drawing on p. 53). If the second cut is not a perfect 90°, then I'll fiddle with the fence until the cut is perfect before screwing it into position permanently. Time spent getting the fence right is time well-spent. If, for aesthetic reasons, you want the rear of the base to be flush with the fence, you can trim the base flush with a bearing-guided, flush-trimming router bit. Either way, the performance of the crosscut box will be unaffected.

Anything from a small wooden handscrew to a fancy commercially made stop will work as a stop block for this fence. A self-stick ruler can be added to the fence or table.

A pivoting fence

I wanted a pivoting fence for making angled cuts, but I also wanted to be able to remove the fence quickly when I need to cut wide

From *Fine Woodworking* (July 1994) 107:40-43

Quick, accurate tenons, even in large boards, *are easy with the author's hinged tenoning jig (right). A hold-down clamp grabs the workpiece securely and accommodates almost any size workpiece. The big footprint of the tenoning jig's base anchors it securely to the base of the crosscut jig below. The jig is also useful for cutting long miters and angled tenons.*

Setting angles accurately *can be done quickly with a miter square or a bevel square (below). By setting the angle both fore and aft in the tenoning jig, you can be sure the angle will be true across the face of the jig.*

boards. I accomplished this first by setting a T-nut for the pivot point into the underside of the jig's base about 6 in. forward of the fixed fence. Then I routed an arc-shaped track for a carriage bolt at the end of the fence (see the drawing on p. 53). The arc runs from 0° to a bit more than 45°, and there's a plunge-routed hole just below the 0° point through which the carriage-bolt assembly can be lifted out to remove the fence. I marked two common angles (22½° and 45°) onto the jig for quick reference using a large protractor and transferring that angle to a bevel square and then to the plywood. These angles can also be checked and fine-tuned by cutting them, setting the resulting blocks together and checking for 90° with an accurate square.

A slotted screw and washer secure the fence at its pivot point but allow the fence to move, and a wing nut (with washer) fixes the angle of the fence at its outboard end. As with the fixed fence, a stop block may be as simple or sophisticated as you like.

An adjustable tenoning jig

A simple hinged jig that uses the rear fence as a reference surface will allow you to cut both regular and angled tenons, rabbets and angled edges accurately and without too much fuss. I built this jig also from Baltic-birch plywood. I crosscut it in the basic jig and routed the slots in it on my router table.

To attach the hinges accurately, I indexed both halves against the fixed rear fence, set a length of piano hinge in place and used a small carpenter's square to align the hinges (see the bottom right photo on the facing page). Then I drilled screw holes using the Vix bit and screwed the hinge on.

A small shopmade (turned) knob at the end of a carriage bolt secures the tenoning jig to a T-nut in the underside of the crosscut box's base. The fixed rear fence ensures that the face of the tenoning jig stays parallel to the blade. Two brass lid supports hold a set angle securely (see the photo at left). And a hold-down clamp travels in a slot in the upper portion of the jig, allowing me to hold almost any size workpiece securely (see the photo at right).

Corner-slotting jig

Attaching directly to the tenoning jig, the corner-slotting jig is easy to build and simple to use. I screwed two scrap boards to a backboard to form a 90° carriage positioned at 45° to the base of the crosscut box (see the drawing). I cut a brace to fit up a few inches from the corner of the 90° carriage and across whatever it is I'm slotting. A hole through the backboard permits a hold-down clamp to bear upon the brace, distributing the pressure of the clamp.

In use, I slide the workpiece into place, then the brace and then I tighten the clamp. The jig feels solid and works well. □

Ken Picou is a designer and woodworker in Austin, Texas.

Shopmade Tablesaw Guards

Building safety into your jigs

by Sandor Nagyszalanczy

***Safety without sacrifice**—A Plexiglas shield keeps hands safely away from the blade without compromising visibility on the author's box-joint jig.*

"Blade guard removed for photo clarity." How many times have you been watching a home-improvement show or woodworking video and seen those words appear across the bottom of the television screen? Well, I want to know: *what* blade guard? In almost all the cases I've seen, a stock tablesaw guard wouldn't have worked in the applications shown.

What's a woodworker to do? Must we continually expose ourselves to unreasonable risks when we perform operations that require removal of the tablesaw's standard blade guard—jobs like sawing tenons, cutting box joints and cove cutting? I suppose we can hope our luck holds out, or we can wait for some kind of sensational all-purpose saw guard to hit the market. But I advocate another alternative: to design safer tablesaw jigs and setups by adding guards and safety devices that prevent accidental contact with the sawblade. I think any woodworker bright enough to design innovative jigs for complicated woodworking tasks could make those same jigs a lot safer without investing too much extra time or material. After all, how much is a finger worth?

In this article, I'll show you some of my solutions for making common tablesaw jigs and setups much safer. One thing I aim for in modifying my jigs is to reduce the degree to which safety relies on judgment. It's a given that, as you work, especially at repetitive tasks, there will be times when your attention flags or is diverted. A safe jig protects you during these lapses. The very best safety feature is one that eliminates the possibility of contacting the blade with anything but the stock. I try to get as close as possible to this ideal in all my jigs.

***Rear guard action**—The simple outrigger behind the box-joint jig lets you complete the cut without exposing the blade.*

In many cases, I've retrofitted existing jigs with guards to show that you don't have to build all new devices to add safety to your woodworking. Because jigs are, by definition, custom-made, the safety measures you take will also have to be individualized. So I haven't tried to cover all the bases here, only to share a few specific solutions and underscore the general idea that safety and guarding features ought to be built into every jig you make.

Clear guards for sliding jigs

Carriages that slide in the tablesaw's miter slots almost always require that the stock guard be removed. Whether you want to use a sliding crosscutting box or a jig for cutting tenons, dovetails or box joints, you can easily retrofit clear blade guards that allow you to see what's going on but keep you from getting cut.

Box-joint-jig guard—I made the guard for my box-joint jig shown in the top photo in about a half-hour from a few scraps of wood and a Plexiglas cutoff purchased from a local plastics store. (Glass shops

Photos: Sandor Nagyszalanczy

and hardware stores often carry clear plastic sheet goods.) The guard is a low box with wood sides and a Plexiglas top that mounts directly over the box-joint jig and provides protection ahead of and after the cut. As an added bonus, I've noticed that it deflects chips and makes dust-collection more efficient.

I made the guard's frame 21 in. wide by 10 in. long, which is wide enough to handle 10-in. drawer sides. I drilled holes in the ⅛-in.-thick Plexiglas sheet so that it could be screwed to the top of the frame (leave the protective paper on the Plexiglas during cutting and drilling to protect it from scratches). When attaching the plastic, I left it about an inch shy of the face of the jig, creating a slot for the workpiece. The 2-in.-high sides provide plenty of clearance between the plastic and the blade. I chamfered and waxed the lower edges of the sides to keep them gliding smoothly. Then I attached the guard to the back side of the box-joint jig with screws through the rear frame member.

To provide blade protection behind the jig, I added a second guard made from a block of wood and a 3-in. by 4-in. piece of Plexiglas, screwed to the underside of the rear frame member (see the bottom photo on the facing page). Even if you don't want to make the entire guard frame, adding a rear guard is an excellent idea. It protects you after the jig has been pushed through the cut when you're reaching over the saw table and are probably the most vulnerable to blade contact.

This type of exit guard is a good addition to any sliding jig. And you can make using it even safer by clamping a stop block to the rip fence or right to the table that will limit the forward travel of the jig—allowing a complete cut through the workpiece but stopping the blade short of the exit guard's rear block.

Tenoning-jig guard—Protecting my hands from the blade involved the addition of three components to my sliding tenoning jig: a clear plastic shield ahead of the cut, an exit block to cover the blade behind the cut and a hand rest to prevent my left hand, which holds the workpiece against the jig, from sliding down into harm's way, as shown in the photo at right. The clear shield is nothing more than a 10-in.-long, 2½-in.-wide piece of ⅛-in.-thick Plexiglas screwed to the edge of a wood strip. This strip mounts to the face of the tenoning jig via slotted holes I made using a straight bit in the plunge router. The slotted holes allow me to shift the shield in or out depending on the width of the workpiece. I glued and screwed a 2½x2x1½ wood exit block to the back of the jig. I used a brass screw just in case it's accidentally hit by one of the two sawblades used during tenoning. A larger block would provide more protection, but as long as you use the jig in conjunction with a stop block, this size is fine. The final component, the hand rest, is a 4x2x1½ block glued to the edge of the tenoning jig's fence. You could position this block higher, if you find it more comfortable.

Crosscut-box guard—A shopmade sliding crosscut box that rides in the tablesaw's miter slots is great for trimming and crosscutting long boards or moldings. And adding a guard is the perfect way to make this sliding jig safer to use. The guard that I made for my crosscut box, as shown in the top photo on p. 58, is basically an inverted U-shaped channel that rests on top of the stock over the line of cut, preventing hands from reaching into the blade. This design is very similar to the clear plastic guard that Kelly Mehler built in his article in *FWW* #89, except that mine was made as a retrofit and has wood sides—I don't miss being able to look through the sides of the guard.

I started building the guard by cutting two 2¼-in.-wide, ⅜-in.-thick wood sides and a 3½-in.-wide, ⅛-in.-thick Plexiglas top, all slightly shorter than the front-to-back dimension inside my crosscut box. I then nailed sides and top together with #16 brass escutcheon pins through holes drilled in the plastic. Because the guard was retrofitted to my crosscut box, I couldn't cut grooves for the ends of the guard to slide in, as in Mehler's design. But for a smaller (12-in. capacity) box like mine, two narrow guide strips tacked on the inside of the box's front support are adequate to keep the guard in place and let it ride up and down. Chamfering and

Untouchable tenoning jig*—An adjustable Plexiglas blade guard and a hand rest combine to keep your exposure to the blade near zero on this tenoning jig. The block that's clamped to the rip fence provides a positive stop and prevents the blade from cutting through the exit block at the back of the jig.*

From *Fine Woodworking* (January 1994) 104:56-59

Crosscuts safe and simple*—A three-sided box over the line of the cut reduces the chance of accidental blade contact on the author's crosscut jig. The box, with ⅜-in. wood sides and a ⅛-in. Plexiglas top, is held in place at one end by two cleats and rides up and down between them. An exit block guards the blade at the end of the cut.*

rounding the ends and edges of the wood sides makes the guard slide up and down easily. To shield the blade where it exits the crosscut box, I added a rear guard that is a variation on the one for the box-joint jig described previously. In this case, I simply glued and screwed on a wood block to sheathe the blade.

Sliding miter-carriage guard—Many woodworkers like to cut miters on the ends of moldings, picture frames and other trim using a carriage with twin 45° fences, which slides in the tablesaw's miter-gauge slots. When you use this type of jig, you hold the workpiece against the fence during the cut, and your fingers often come close to the blade. And as you finish the cut, the blade exits between the fences, not far from where your thumbs are wrapped over the top of the fences. It's an operation that begs for a guard.

To add protection to my sliding miter jig shown in the bottom photo on the facing page, I cut a triangular block from some scrap 2x4 I had around the shop and glued and screwed it to the jig's baseplate just behind the intersection of the fences. This block acts as an exit guard and a mounting surface for a clear blade guard. The back end of this blade guard, a 5-in. by 12-in. piece of ⅛-in. Plexiglas, is screwed to the top of the block, and the front end is screwed to a wood strip nailed to the miter jig's front cross support. To complete the safety treatment, I clamp a stop block to the saw table to prevent the blade from cutting through the exit block.

Two resawing guards

Probably one of the most dangerous operations to perform on an unguarded tablesaw is resawing, for two reasons: First, the blade is usually raised to or near its full height. If there's a slipup, you are exposed to more harm than with any other tablesaw operation. Second, there is maximum surface area contact between the wood and the blade. If the wood distorts and binds between the fence and blade (or the kerf closes up and pinches the blade), the workpiece is kicked back with the full force of the saw. These are two excellent reasons to invest a few minutes and a couple of pieces of wood to protect yourself against disaster.

I've come up with a pair of guarding devices for resawing. Both are simple, but effective. These jigs serve two purposes: They keep the board upright during the cut, and they keep your hands from coming anywhere near the blade.

The first is a clamp-on guard, as shown in the photo at right. It consists of a 12-in.-long block of 4x4 lumber with a 2x2 stick screwed to one side. At 3½ in., the 4x4 is thicker than the depth of cut of most 10-in. tablesaws (if your sawblade rises higher, use a thicker block). The block is positioned over the throat plate, just far enough to the left of the blade to allow the stock to feed past. Because the resawn stock will have to be planed anyway, you can set the guard for a fractionally loose fit to account for the distortion caused when the workpiece is cut. The 2x2 stick should be made long enough to center the 4x4 with respect to the blade arbor.

To use the clamp-on resaw guard, set the rip fence, lower the blade into the table and put a piece of stock in place above the blade. Then position the block so it's over the throat plate and snugged up to the workpiece. Secure the end of the stick to the saw table with a C-clamp.

If you do a lot of resawing, you might want to make the second style of guard, which incorporates a dedicated throat plate. On this device, the wood block is attached directly to a replacement throat plate. In addition to providing protection like the clamp-on guard, this version enables you to raise the sawblade through the blank plate for a close fit that supports narrow workpieces right next to the blade. And it prevents the leading edge of the work from hanging up.

Make the replacement throat plate from plywood, particleboard or Masonite that's the same thickness as the original plate. The easiest way I've found to shape the new plate is to use the factory throat plate as a template. I cut out a slightly oversized blank on the bandsaw, attach the factory plate to it with Scotch brand 924 Adhesive Transfer Tape (available in ½ in. and ¾ in. widths from University Products, 517 Main St., Holyoke, Mass. 01041; 800-628-1912) and then trim the new one to size using a piloted, flush-trimming router bit. Once the new plate fits snugly in your saw, screw on the block from below. I keep a couple

Resawing reconsidered*—A chunk of 4x4 screwed to a stick is all that it takes to keep the stock vertical and the blade safely hidden while resawing. If you resaw often, you can screw the guard block directly to a dedicated throat plate.*

of these dedicated throat plates handy—one for resawing 4/4 stock and one for 8/4. You can cut slots instead of holes for the screws through the replacement blank to permit adjustment for resawing boards of various thicknesses.

When working with either style of resaw guard, use a push stick to feed the end of the stock through the gap between block and fence—even if the blade is buried in the wood. If resawing must be done in two passes, set the blade height to slightly less than half the width of the board. The board is easily snapped apart after the second pass, and the small unsawn strip down the center of each resawn half can then be planed off. Incidentally, you can also use a similar guard—with a block that's not as high—when ripping narrow strips to width.

***Wide featherboards are excellent for coving**—They exert downward pressure over the cutting area while keeping hands from coming near the blade.*

Hold-down cove-cutting guard

Passing your hands directly over the blade is dangerous, even if the blade is buried in a thick workpiece—the stock might be kicked back, suddenly exposing the blade. In tablesaw cove cutting, you have to keep constant downward pressure on the workpiece to get good results, so this danger is always present.

My cove-cutting guard, as shown in the photo above, is attached directly to a clamp-on fence, which guides the workpiece across the blade. The guard employs a featherboard-style hold-down over the blade. The hold-down prevents fingers from getting near the blade while keeping the stock flat on the table. And because the hold-down is firmly positioned, it does a better job of flattening the stock than your hands can. The only thing better than a guard like this is a power feeder, which will keep the stock flat on the table and your hands safely away from the blade while feeding the piece for you.

I made the cove-cutting fence from straight-grained stock; I used a 1¾-in.-wide, 1½-in.-thick piece of Douglas fir. A block of wood 1¾x3x4 is screwed to the top of this fence. Its position along the fence varies depending on the angle of the fence, which is determined by the desired cove profile (for more on cove cutting, see "Coves Cut on the Tablesaw," *FWW* #102, p. 82). I cut the featherboard from a 4½-in.-long, 3-in.-wide, 2-in.-thick block and cut the feathers on the bandsaw, making each one about ³⁄₃₂ in. thick. Then I attached the featherboard to the fence block with a ⅜-in.-dia. carriage bolt.

***Miter shield**—A triangular piece of 2x stock serves as an exit block as well as a mounting surface for the Plexiglas blade guard on this sliding miter-carriage jig.*

To use the device, clamp the fence to the saw table to the right of the sawblade with the guard centered over the blade. With the sawblade lowered into the table, place the workpiece under the featherboard. Pivot the featherboard until it exerts enough pressure on the piece to press it flat, but not so much that the workpiece is difficult to feed. Depending on the thickness of the work, you may have to relocate the hole for the carriage bolt in the fence block. Finally, clamp a secondary fence to the saw table to keep the work from wandering away from the main fence during cove cutting. As you make each pass over the blade (the blade should only cut about ¹⁄₁₆ in. deep each pass), use the next workpiece or a piece of scrapwood the same width as the workpiece to push the end of the work under the featherboard. □

Sandor Nagyszalanczy is a contributing editor to Fine Woodworking *and a writer, musician and furniture designer/craftsman in Santa Cruz, Calif.*

Controlling Wood Dust

Four thrifty shop-built devices use cabinets, filters and vacuums for collection

by Alec Waters

Wood dust is annoying. Whether you're trying to apply a flawless finish, maintain machinery or keep your shop fire-safe and clean, sawdust and wood chips are a nuisance. More important, though, is the damage that wood dust can do to your body (see *FWW* #83, p. 72). Although humans have fairly effective filtering mechanisms in their noses and lungs, the dust present in woodshop concentrations (see the photo above) can be toxic, and even carcinogenic. In 1989, the Occupational Safety and Health Administration (OSHA) established industry guidelines for dust. For hardwood and softwood dust, the permissible exposure level (PEL) of respirable dust is 5 mg. per cubic meter of air. The total allowable dust is 15 mg./meter. So what does this mean to the average woodworker? It means dust collection and air filtration in the shop are more important now than ever.

Luckily, there are hoards of dust-sucking machines available commercially. For the modest needs of carvers, there are lap-top models (In-Lap Dust Collection Systems, P.O. Box 081576, Racine, Wis. 53408; 414-633-8899). And for the high volumes of dust in production shops, there are cyclone separators (for more on these, see *FWW* #100, p. 76 and *FWW* #103, p. 34). However, for many small-shop owners, the big price tag and size of the manufactured and high-end collectors are deterrents. That's why many woodworkers have come up with their own dust-controlling ideas. Over the last year, *Fine Woodworking* has gathered an as-

Photos except where noted: Alec Waters

Dealing with two kinds of dust—*Wood dust can be best handled by breaking it into two components: heavier chips or shavings and finer dust. Hence, the need for two-stage collectors, which settle out larger particles before the air stream enters the impeller to deposit the finer dust into a filter bag. Generally, machines with cutterheads, like this planer, produce more chips while saws, sanders and routers produce more dust.*

sortment of shopmade systems submitted by readers. Some units are frugally cobbled together from scrapwood and spare parts; others resemble professionally built machines. I've picked out a sampling of units, both simple and involved, to show what home-brewed ingenuity and resourcefulness can yield. But before I share the designs, I'll discuss general dust-handling strategies.

Collectors, vacuums and filters—oh, my!

Sawdust actually consists of a range of particle sizes. Both single and two-stage dust extractors, so-called source-capture collectors, use impellers (a rotor with fan-like blades) that propel air and dust through ducts to a storage container, usually a bag. But two-stage units take advantage of different particle sizes. They divide heavier chips from lighter dust before the mixture reaches the impeller. The first stage relies on gravity to cause heavier particles to fall into a drop box (usually a barrel or bin). The lighter dust continues on to be collected in the bag. But don't breathe easy yet.

Most dust-collection systems capture from 50 to 90% of the dust. Also, the bags themselves catch dust only so fine. As one reader, Daryl Rosenblatt, says: "Dust collection is a philosophy. No single collector will get it all." And it's the tiny particles (those under 10 microns are respirable) that are so damaging when inhaled, especially to those who suffer allergies. (For a chart of toxic woods, send $2.50 to the Center for Safety in the Arts, 5 Beekman St., Suite 820, New York, N.Y. 10038; 212-227-6220). Depending on the person, the exposure level and the wood species, symptoms can range from eye, nasal and skin irritation to respiratory and cardiac problems. That's where the free-hanging filtration units come into play. Some operate electrostatically (charging the dust particles so they can be removed), and others are fan-powered. Both types use filters to manage the dust. But the only sure way to protect your lungs is to use a fresh-air-supplied respirator that has a proper-fitting mask. At the least, you should use a dust mask. In fact, after much tribulation trying to make his home shop dust-free, Rosenblatt now advocates a four-system approach: a dust collector, a shop vacuum, an air-filtration unit and a respirator.

The shortcomings of conventional dust collectors have prodded other solutions. Because of severe wood allergies, John Timby, a New Mexico woodworker and retired design engineer, developed a two-bag (one is impervious) extraction unit called a "depression chamber" that keeps dust from re-entering the shop. It's designed to remove all micron-sized and under particles. Timby also offers a pair of video tapes for $60, which explore this unit and ways to hook up dust collectors to stationary machines. For more information, write to John Timby, P.O. Box 1904, Deming, N.M. 88031.

Designing your own system

Now that you know how dust behaves, you have to decide what is best for your shop. When designing a collection system, you will want to properly size its motor (in hp), air handling rate (in cubic feet per minute) and ductwork (in diameter, length and junctions) around your machine requirements (for more on this, see Roy Berendsohn's article in *Fine Woodworking* #67, p. 70). Be aware, too, that universal motors, commonly found in vacuums, won't hold up as well as induction motors most often used in commercial dust collectors.

When it's time to build your system, be sure to eliminate fire and other hazards. For example, ground the duct work to dissipate any static charge, and try to select an impeller material that won't conduct sparks. Also, see that bags have adequate capacity, and clean them often, and see that filters are fine mesh, but won't clog. For air-filtration units, avoid creating an air-flow pattern that will blow across your face. Finally, make sure that there's enough fresh air coming in the shop to replace what's being exhausted.

Designing your own system quickly leads you to where the dust is collected. And here is where the stories on the following pages will help. In all cases, the units were inexpensive to build using readily available materials like plywood. The first unit is a cabinet that collects dust and chips from the most demanding machine in the shop: the planer. The second is a ceiling-mounted air-filtration box, which hangs out of the way and runs quietly. The third unit is a portable dust-collection box, which has an easy-to-clean bin and is fed by a standard barrel-top collector. The fourth unit is actually a mobile stand with an adjustable collector hood that is powered by an ordinary shop vacuum. Just like building a furniture project, the nice thing about these shop-built collectors is you can mix and match features to fit your needs. If you're still not happy with the results (or if your spouse incessantly complains of wood dust and noise), you can always go back to making shavings with hand tools. □

Alec Waters is an assistant editor for Fine Woodworking.

Portable-planer chip collector

Vacuum motor in base cabinet sucks up shavings and dust

by George M. Fulton

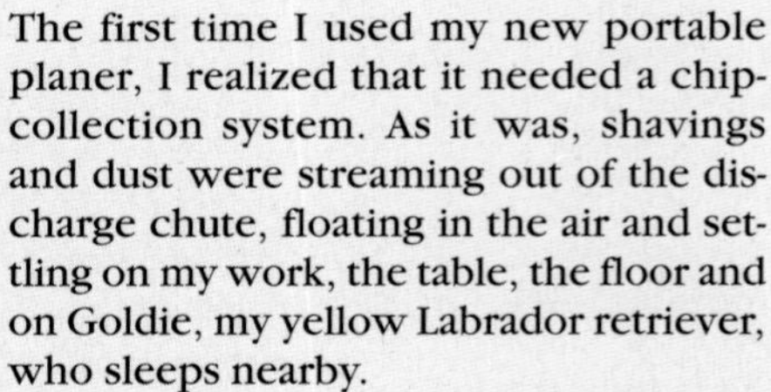

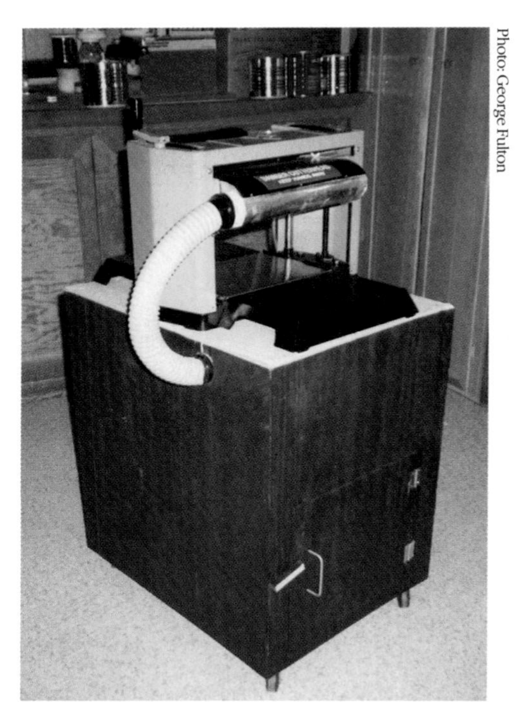

Photo: George Fulton

Planer chips conquered*—To tame his biggest chip maker, George Fulton took a discarded vacuum motor and built this combination planer stand and dust cabinet. By attaching a wand to the flexible hose, he can vacuum up leftover dust.*

The first time I used my new portable planer, I realized that it needed a chip-collection system. As it was, shavings and dust were streaming out of the discharge chute, floating in the air and settling on my work, the table, the floor and on Goldie, my yellow Labrador retriever, who sleeps nearby.

I decided to make a chip-collector cabinet that would remove dust and serve as the base for the machine (see the photo at left). The cabinet had to be compact, connect easily to my Delta planer without substantial modification, and it had to be inexpensive and easy to build. It also had to be stable, like a stand, but mobile so I could wheel the planer out of the way.

Construction: You should be able to adapt the cabinet to any machine by slightly modifying the dimensions or construction shown in the drawing on the facing page. Basically, the cabinet consists of a frame boxed with plywood, a vacuum compartment and top made of plywood, a discarded vacuum motor and a plastic cat litter tray. On the infeed side of the cabinet, I mounted a screened vent to cool the motor, and I built a drawer to hold miscellaneous adapters and tools. On the side of the cabinet, I made a vacuum inlet, and on the outfeed end, I added a clean-out door. I fashioned a dust hood (the manufacturer didn't offer one at the time) out of sheet metal, which mounts to the planer and provides a way to connect the dust-inlet hose to the cabinet.

Dust hood: I formed the dust hood out of .017-in.-thick (27 gage) galvanized sheet metal, making sure that the hood wouldn't interfere with the chip deflector or a workpiece. I riveted the hood to the guard, and then I installed 2-in.-dia. flexible hose, which was compatible with the PVC pipe fittings I had. If you want to hook the cabinet up to a standard collector, you'll probably want to use 3- or 4-in. hose and fittings. Because the planer's thicknessing range is achieved by raising and lowering the cutterhead, I used flexible hose. That also lets me easily disconnect it from the hood, pop on a standard vacuum pick-up wand and clean up dust around the planer.

Cabinet: I constructed the frame, as shown in the drawing on the facing page, using 2x4s and 2x2s. For the box sides, top and vacuum chamber, I used ½-in. and ⅜-in. plywood. To make cleanup easier, I adhered plastic laminate to the cabinet top. I mounted a discarded vacuum-cleaner motor in the exhaust chamber using aluminum angle brackets. I grounded the motor housing using a lug terminal and the green wire of the motor cable. After I installed the vent and drawer on the infeed end, I added a partition between them, so the air is exhausted through the vent and the opening in the floor of the box.

Although the vacuum compartment's seams were tight, I applied a bead of caulk all around the interior corners and inlet box joints. Because the clean-out door had potential to leak air, I surround-

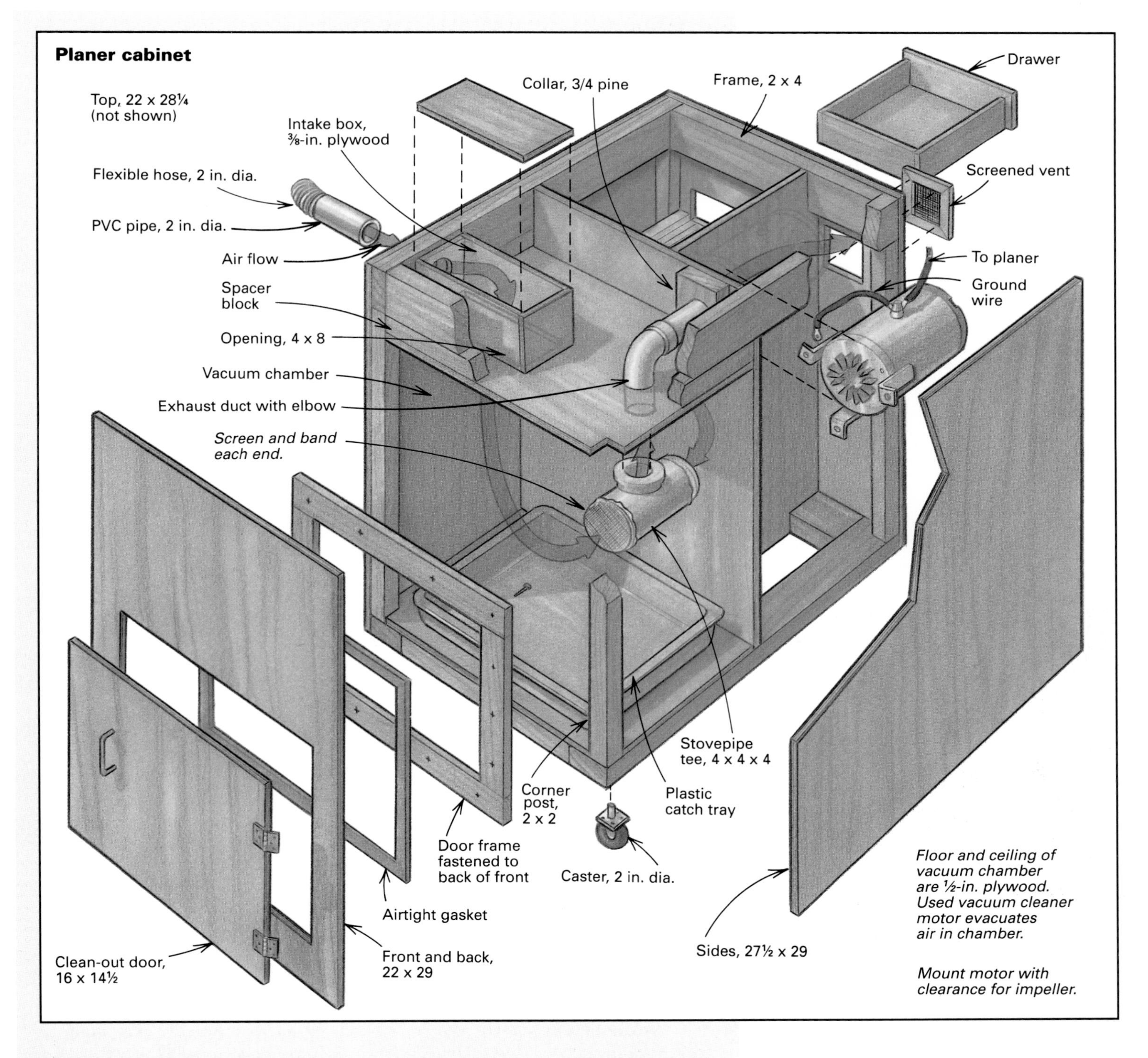

ed the opening's inner frame with a ¼-in. by ¾-in. weather-seal gasket. A pivoting latch compresses door to gasket.

Duct work: The exhaust duct consists of a pine collar, 2-in. PVC pipe and elbow, and a 4-in. tee. I formed a section of aluminum window screen over the two open ends of the tee and secured them with rubber bands. You could cover the screen with nylon stocking to further filter dust. For the inlet duct, I used 2-in. PVC, flexible hose and threaded coupling (see the photo on the facing page).

Wiring and final details: To allow the vacuum to run after the planer is off, I installed a toggle switch next to the cutterhead switch and wired it to the motor. After I secured four furniture casters to the bottom frame of the cabinet, I mounted the planer to the cabinet with bolts and T-nuts. Finally, I placed a plastic waste tray in the vacuum chamber under the inlet box to gather the lion's share of shavings.

Now my dog Goldie dozes fairly contentedly, although she is probably wondering if something can be done about all the noise. □

George Fulton is a retired electrical engineer and a hobbyist woodworker in Arnold, Md.

Drawings: Bob La Pointe

Shop air-filtration box

Get additional protection from fine dust

by Jim Whetstone

Photo: Jim Whetstone

Ready to filter shop air*—After installing a fine-mesh, filter in his ceiling-hung filtration box, Jim Whetstone can breathe easier. Although he owns a shop vacuum and dust collector already, he wanted additional protection from finer dust.*

I was convinced that my shop's exhaust system was not that efficient. I suppose it was removing fumes and radon gas adequately, but I felt it needed to do a better job of removing dust from the air. So I decided to build an air-filtration box. I made it out of plywood, a fan, a timer and a household furnace filter, which catches the dust. The box hangs from the ceiling: still accessible but out of the way (see the photo at left).

Design and materials: For the box, I used about one-quarter of a 4x8 sheet of ¾-in. birch plywood. The fan is a 9-in. axial fan motor ($55) from Grainger (call 800-473-3473 for the nearest location). I bought a 30-minute mechanical timer (instead of an on/off switch), so I could leave the shop with the fan running. Originally, I used a 12x12 fiberglass furnace filter, which worked okay. But lately, I've been using a finer-mesh synthetic filter, which costs about $1.20 (from American Air Filter Co. 33 Industrial Road, Elizabethtown, Penn. 17022; 717-367-5060). A filter, which arrests 85% of particles in the 3 to 5 micron size, lasts about three weeks if I'm using my machinery heavily. (Be sure to write the installation date on new filters.) I picked up the rest of the hardware and wiring (see the drawing) at my hardware store.

When building the box, I called Grainger to see about the fan-spacing requirements. They advised that a 6-in. space between the filter and the fan would be fine. I made the box's top 2¾ in. longer than the box so that I'd have a mounting surface. I routed rabbets to receive the fan mounting piece and the filter. Before I glued up the box, I drilled all its holes, including the ones for the electrical box.

A simple open-panel door holds the filter in place. I used ¾-in. pine for the door's half-lapped frame and to help direct dust into the box, I chamfered the door's inside edges. A pair of small hinges and a hook-and-eye catch secure the door to the box.

On the back of the box, I cut a 9-in.-dia. hole for the fan using my jigsaw. After I screwed the fan to the back, I dry-assembled all the parts and turned the unit on. Everything worked properly, so I stripped the hardware, sanded the box and then painted and urethaned it to match my other shop cabinets.

Air-filtration box

Top, 11½ x 19, mounts to ceiling.

Countersink underside of hole.

To power

Stop, ⅜ x ¾, each side

Timer switch

Electrical workbox

Back, 12¾ x 12, has 9-in.-dia. fan opening.

Sandwiched ½ x ½ hardware cloth

Rabbeted for back

Distance from rear of filter to fan is 6 in.

Side, 11½ x 12⅜

Bottom, 11½ x 12

Rabbeted for filter

Filter, 12 x 12

Door frame, ¾ x 1¼ pine

Box is made of ¾-in. birch plywood.

Assembly and mounting: When putting the box back together, I added lock washers while mounting the fan to the back. For safety reasons, I sandwiched a piece of ½-in. by ½-in. metal hardware cloth between the fan and plywood. Next I installed the electrical box and wired the timer to the fan. I located the air-filtration box over my bench where there's good head room and a nearby duplex ceiling receptacle. I drilled holes in the box's top 16 in. on center to match my ceiling joist spacing. Finally, I screwed the air-filtration box in place, inserting ¼-in.-thick wood spacers behind the screws, so the unit would hang below the ceiling slightly.

I've been using the 560-cu.-ft.-per-minute filtration box off and on now for a year and have noticed the air is definitely less dusty, though I still use a respirator for certain work. Also,the noise level is quite acceptable (47 dB). I can still hear the radio or television. □

Jim Whetstone has been working wood in New Cumberland, Pa., for more than 25 years.

Dust-collection box

Replacement for conventional drum makes clean-out easy

by Jack Minassian

I built a dust-collection box to replace the 55-gal. steel drum that my Delta dust-collection unit is designed for. The box has a drawer, which lets me clean out dust without having to remove the heavy motor from the drum. The cabinet is 3/8-in. Baltic-birch plywood with exterior poplar strips that protect the plywood edges and allow easy assembly. The drawer, which has waxed oak runners and guides, is made of 1/4-in. plywood.

The photo shown at right shows how the box is made. Most of it assembles easily, but a few details are worth noting. When cutting the 22-in.-dia. opening in the top, use a sabersaw fitted with a radius strip, and pivot the saw like a compass. The resulting circular cutout can be used as a form for the support ring. I made my 1-in.-wide ring by laminating 7-ft. strips of 3/32-in. by 1 1/8-in. poplar with scarf-joint ends.

To clamp the ring, I used 16-gauge by 3/4-in. nails with square wooden pads under the heads. To remove the nails, I pried under the pads. Next I placed the ring over the form. Then, using a center pivot and a sanding disc on my tablesaw, I sanded the ring perfectly round. With the blade back in the saw, I held the ring vertical and rotated it to rip the ring to its correct thickness.

The inside of the door has a gasket made of 1/8-in.-thick self-sticking neoprene, which will compress to about 1/16 in. I hung the door on a 1 1/2-in.-wide piano hinge, and then I installed blocks to the cabinet sides to mount two buckle-hasp latches (made by Brainerd Manufacturing Co.), which are available at most hardware stores.

You can round the edges of the poplar trim either before, or once everything is assembled, using a 1/4-in. roundover bit and a router. After you determine the position of your collection unit, shape the support ring for a snug fit. Finish the box as you like (I painted mine Powermatic-machinery green), and then fasten 2-in., 90-lb. swivel casters to the bottom. Finally, install your collector on top. □

Jack Minassian is a retired architect. For a detailed construction drawing, send $6 to Jack Minassian, 15-20 201 St., Bayside, N.Y. 11360.

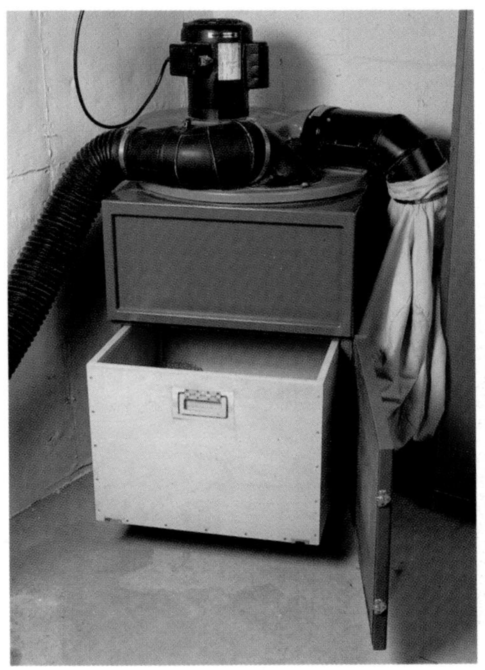

***Mobile box improves collector**—By replacing his dust collector's 55-gal. drum with a cabinet, Jack Minassian can wheel the unit to any machine in the shop. Opening the door reveals a file-cabinet-like drawer, which can be readily emptied.*

Mobile stand with intake hood

Versatile setup handles a variety of sanding chores

by Gregor Jakob

Sanding dust is an ever-present problem in my woodshop. I've used face masks and left the windows open; then I designed a dust-collector stand that connects to my shop vacuum. The setup works for benchtop sanding (see the photo) and for my stationary drum sander.

To make the stand, I used plywood, pine, melamine, arborite laminate and metal stove pipe. The base has casters and a telescoping column, which provides height adjustment. A pivoting oak head allows the funnel-shaped hood to swivel and tilt. The hood's adapter tube fits my 2-in.-dia. vacuum line. □

Gregor Jakob is a technology teacher in Mississauga, Ont., Canada.

Photo: Gregor Jakob

***Dust-removing helper**—When sanding, Jakob rolls up this hooded stand and connects it to his shop vacuum.*

Sliding Table Simplifies Mortising

Heavy-duty drawer slides for precise alignment, easy action

by Mac Campbell

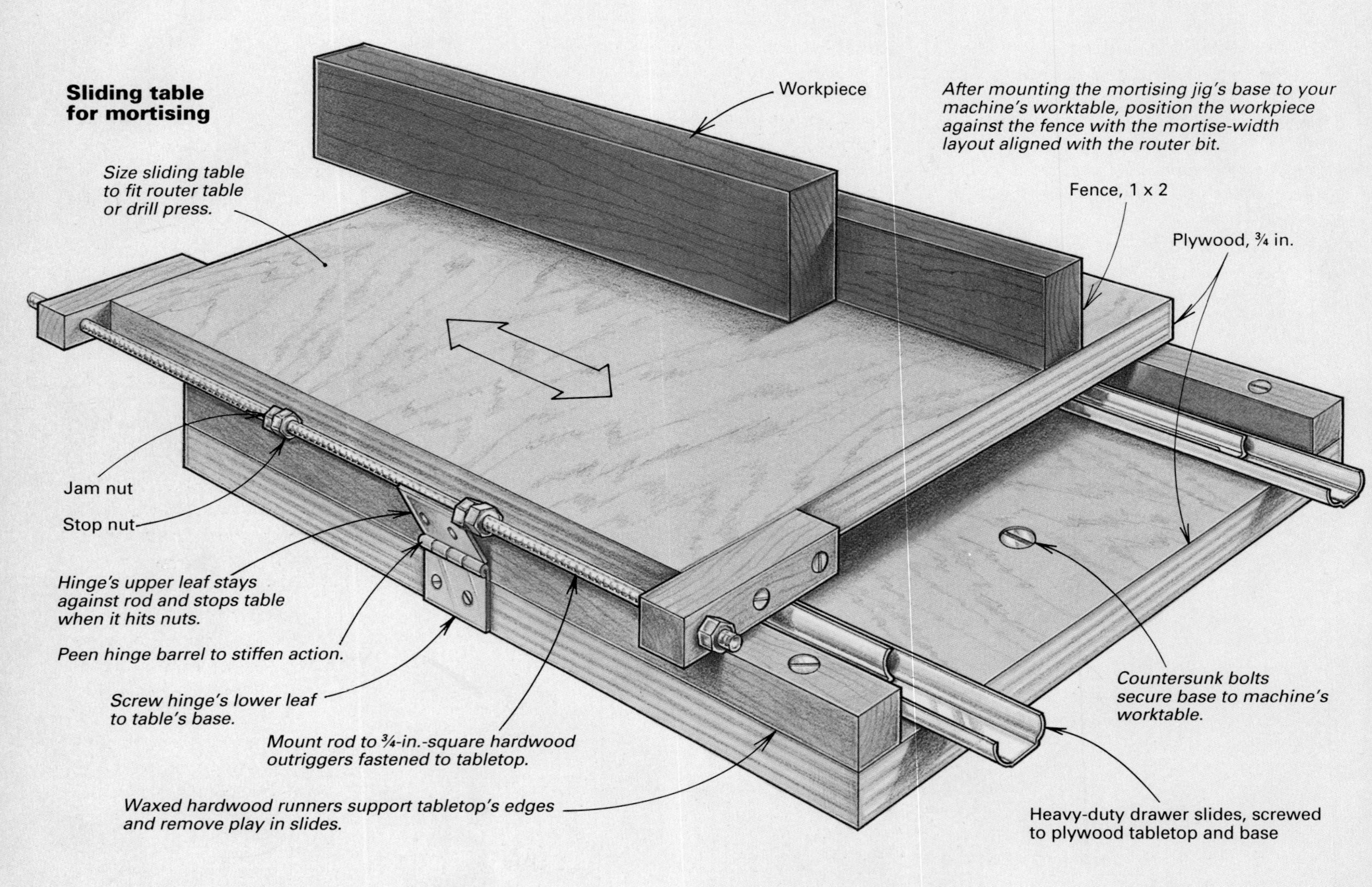

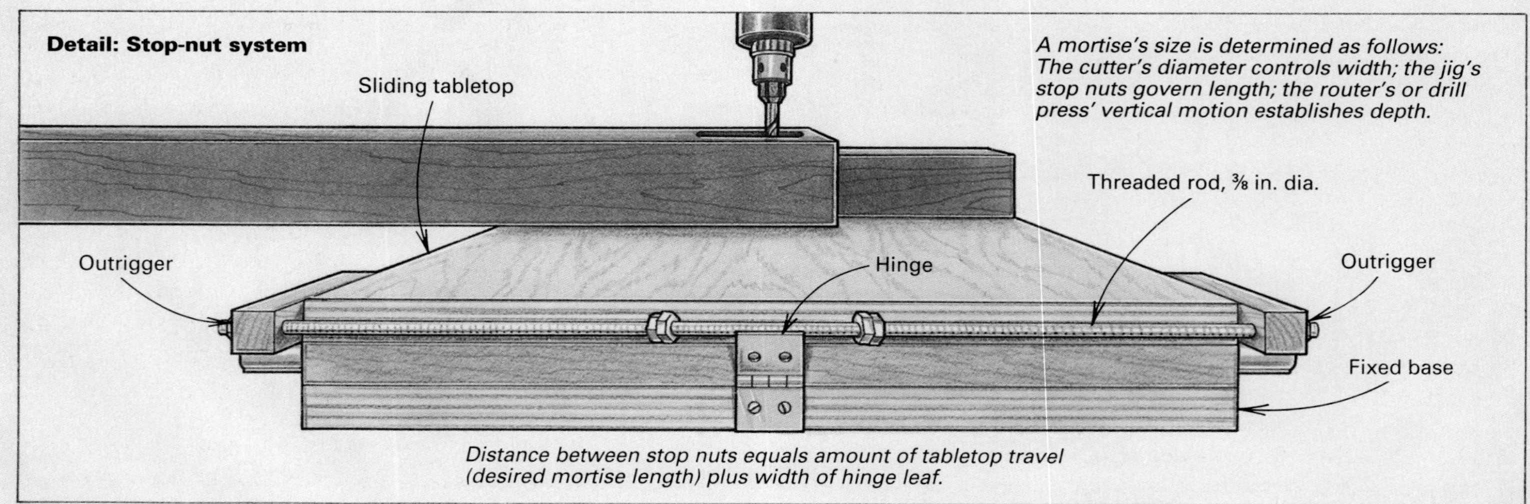

Photo: Alec Waters; drawing: David Dann

When I need to cut lots of mortises, I like to use a tool that does a consistent job without requiring an involved setup each time. A few years ago, I had to mortise a sizable run of custom chair components, so I picked up a small, used overarm router. To make mortising more efficient, I fitted the machine with a sliding table (see the photo below), which was quick to build using plywood and standard hardware I had on hand. Adjustable stops make setting the table's horizontal movement straightforward, and the pedal-fed router makes vertical (mortise depth) setup and plunging a cinch.

Rockwell no longer manufactures the machine I have, but similar tools that have a rigid arm supporting a router at 90° to a height-adjustable table are still made. Even without an overhead router, the sliding table can be mounted on a drill press or used for horizontal routing. In addition, the table's indexing (stop) system is especially well-suited to hollow-chisel mortising.

Assembling the table

The sliding table consists of upper and lower plywood pieces connected by pairs of drawer slides and hardwood runners, as shown in the drawing. The table size depends on the tool you are mounting it to and on the length of the drawer slides you're using. The slide mechanisms let the tabletop travel laterally over the base, which is bolted to the tool. The larger the table (mine is 20 in. long), the more stable the setup will be. A piece of 1x2 secured across the width of the tabletop serves as a fence.

The drawer slides are the heavy-duty variety intended for file cabinets. They are equipped with a metal track that's designed to be mounted to a cabinet side and a similar track for the drawer side. A ball-bearing carriage runs between the two tracks. My slides are 20 inches or so long and allow about 25 inches of travel. These slides have virtually no play and don't depend on gravity to keep everything aligned. A little judicious tinkering will remove the stop on each slide that prevents the drawer from over-closing, allowing the mortising jig to extend in both directions.

The hardwood runners support the tabletop's edges and eliminate any play that might develop at the extremes of travel when there is a relatively short length of the bearing carriage between tracks. The runners should be thick enough to fill the gap between the sliding table and the base. Waxing the runners lets the jig move smoothly.

***Ideal for mortising, this shopmade sliding table** advances a mahogany stile under a fluted stagger-tooth bit chucked in an overarm router. Author Mac Campbell secures the workpiece to the fence with a block of scrap and a C-clamp.*

Installing adjustable stops

A length of ⅜-in.-dia. threaded rod installed along the front of the upper table (see the drawing detail) is the basis for the table's stop system. I bolted the rod to hardwood outriggers on the ends of the table. Nuts threaded onto the rod serve as stops that regulate table movement. Several pairs of nuts act as multiple stops; the second nut in each pair (a jam nut) locks the first in place. A hinge screwed to the plywood base bears against the rod and stops the table when its upper loose leaf hits the nuts. The secret for making the hinge stiff enough for the upper leaf to remain upright against the rod is to hammer on the hinge barrel a little.

Mortising on a sliding table

Once everything is assembled and mounted on the overarm router table, I install a straight-cutting bit and then set it to take a fine shaving off the fence itself. This operation trues the fence to the bit regardless of irregularities anywhere along the line. This, in turn, guarantees that the mortises will be parallel to the face of the stock that's clamped to the fence. When mortising with my overarm router, I use a two-flute, stagger-toothed carbide bit, which is available from Furnima Industrial Carbide, P.O. Box 308, Barry's Bay, Ontario, Canada KOJ1B0; (800) 267-0744. I'm told that these fluted bits will only work at speeds of 17,000 rpm and up. Therefore, using them in a drill press is *not* an option.

To mortise a stile, leg or what have you, first lay out the mortise on the stock. It need not be centered; in fact, many applications work better with an off-centered mortise and tenon. A chair rail, for instance, should have the tenon near its outside face, allowing greater penetration into the leg without cutting into the tenon on the second rail, which enters the leg at 90°. After marking the length of the mortise (tenon width) on the appropriate pieces, I locate the center of the layout lines in approximately the correct position on the jig and then clamp the work to the fence. To ensure uniform depth, I preset the machine's depth stops. Because the bit determines the width of the mortise (tenon thickness), cutting mortises is just a matter of aligning the stops with the marks on the stock and routing the slot.

I raise the machine's table to take about ⅛ in. each pass while I slide the jig (with workpiece clamped to the fence) back and forth under the cutter (see the photo). Each mortise takes only 10 or 15 seconds, and changing the stock takes about the same. A series of mortises for 20 frame-and-panel doors takes only a half hour or so, plus, perhaps, five minutes for setup. □

Mac Campbell is studying theology in Halifax, Nova Scotia, Canada. Previously, he ran a custom furnituremaking shop in Harvey Station, New Brunswick.

From *Fine Woodworking* (May 1993) 100:66-67

Fine Furniture for Tools

Tool chest combines storage and convenience while showing off its maker's skills

by Steven Thomas Bunn

***Fit for a showroom**—Because the author doesn't have a display of finished furniture, he built this toolbox to advertise his capabilities to drop-in customers. It also offers lots of convenient storage with 20 removable drawers that can be carried to the workbench.*

I wanted a toolbox that was both visually striking and had a lot of storage space. Appearance was a prime consideration because as a one-man shop, I can't afford to keep finished work around as showpieces, and piles of wood or half-finished parts are not impressive to a drop-in client who isn't familiar with cabinetmaking. I needed a toolbox that, like the journeyman's boxes of old, was an advertisement and demonstration of my capabilities.

I like the European-style toolbox that hangs on the wall with tools hung neatly inside. However, I don't like the large volume of wasted space behind the closed doors. In addition, the sheer number and weight of tools I possess ruled out a box that could be hung on the wall. I like the out-of-sight storage of drawers, similar to a mechanic's toolbox. I also like the idea of grouping similar tools in a single drawer so that I can pull out a drawer of chisels or gouges, set it on my bench and then go to work. Also, drawers keep sawdust and wood chips from accumulating over my tools.

Incorporating drawers meant the cabinet needed to be relatively deep: I calculated about 17 in. deep to be effective. For both design and practical reasons, I decided to put the tool chest on its own stand, as shown in the photo above. The cabinet and stand offer

Photos: Charley Robinson

exceptional storage capacity for fine hand tools at a height that keeps me from having to reach up or bend down to get to anything. But with some slight modifications of the interior storage arrangements, the tool chest could easily house linens, china or electronic equipment. In fact, my tool chest is an interpretation of the Gate's sewing cabinet shown in *Measured Shop Drawings for American Furniture* by Thomas Moser (Sterling Publishing Co., Inc., N.Y.; 1985).

Shop or home furniture? *This tool chest could be equally at home in the parlor with just minor changes to the interior to accommodate china, silver or even stereo equipment.*

Building the carcase

The solid panels of the case top, sides, shelves and bottom are all made of ¾-in.-thick stock, as shown in the drawing on p. 71. After preparing the stock, I routed stopped dadoes into the side panels for the shelves and bottom, guiding the router against a fence. To ensure the dadoes were aligned, I clamped the sides together with the back edges butted against each other. I positioned the fence, squaring it to the front edge of one of the side panels, and clamped it in place.

After the dadoes had been routed into the sides, I joined the top and sides of the carcase with through-dovetails and then dry-assembled the joints. Once satisfied with the fit of the dovetails, I cut a rabbet on the inside back edge of the top and sides for the back. The rabbet in the top was stopped at each end and squared up with a chisel. I then reassembled and glued the dovetails. The bottom was sprung into its dadoes, aligned 1 in. behind the case front to allow for the bottom face frame and screwed into place with glue blocks from underneath, as shown in the drawing.

The shelves were driven in from behind with taps from a dead-blow mallet. The front of the shelves were notched to fit tightly against the case side and hide the dadoes. I left a ⅛-in. gap at the back of all the shelves as a safety measure in case of unequal expansion in the sides and shelves. So if the shelves swell more than the sides, the shelves won't break out the back of the case. Only the bottom was left full depth to provide a place to anchor the board-and-spline back. The interior shelves also stop 3 in. shy of the front to leave room for the tools hung on the inside of each door.

Oversized storage holds big items—*This two-drawer box slides into place between the shelves to hold long items that don't fit in the smaller drawers.*

Slide-in drawer dividers

Four vertical drawer dividers slide into dadoes routed in the two top shelves to form the drawer support system, as shown in the drawing. Before installation, I cut matching dadoes in all four dividers to make the drawer-guide grooves. I followed Tage Frid's advice in *FWW on The Small Workshop*, pp. 18-19 (The Taunton Press) and made a series of grooves at 1¼-in. intervals. The theory is that you could make drawers in 1¼-in. increments for greater storage flexibility. Using Frid's modular system, you could take out two 1¼-in. drawers and replace them with one 2½-in. drawer. I've found this doesn't work in the real world. I'm not about to start making new drawers to replace ones that I already have, and changing the drawers around makes finding tools a guessing game. But beyond that, I'm tired of always being asked, "how come there are more grooves in the dividers than drawers?"

The two outside drawer dividers were installed first. Then I locked the shelves in place with one screw at each end of the shelf, driven through the carcase sides from the outside, as shown in the drawing. The counterbored and plugged screw was centered in the shelf about 1 in. behind the leading edge. The unglued shelf is free to float in its dado behind the locking screw. The two center dividers were added last from the back of the cabinet.

All drawer parts were batched together and cut at the same time. I built the drawers, wherever possible, from wood scrap except the drawer fronts, where I attempted to cut all three fronts in a row from the same cherry board for grain matching. The drawers were half-blind dovetailed at the front and through-dovetailed at the back, as shown in the drawing.

Before final glue-up, I cut the groove for the drawer bottom in the sides and fronts on the tablesaw. The back was trimmed, as shown in the drawing on the facing page, so that the drawer bottom extends past the back and can move with the seasons. Planning the drawer bottom so that just the right amount protrudes lets the bottom act as a stop against the cabinet back when the drawer is closed.

After final fitting of the drawers to their openings, I glued strips to each side. The strips act as drawer runners, and they fit into the grooves cut into the vertical drawer supports.

The two longer drawers below the main drawer section are housed in a separate box, which was an addition that I made later to store tools like rulers and oversized screwdrivers.

***Preventing door sag**—Legs screwed to the doors' lock stiles help support the heavy tools hung on the doors and prevent the hinge screws from pulling out.*

***Supporting heavy loads**—Corner braces reinforce the mortised-and-tenoned rail-to-leg joint and enable this elegant base to support the heavy tool chest.*

Board-and-spline back for seasonal movement

The back is made from seven boards (see the drawing). A ⅛-in.-wide by ½-in.-deep groove was cut into both edges of five of the boards and only one edge of the remaining two boards. The two boards with only one groove were glued on their ungrooved long edges to the case-back rabbet. The remaining boards were screwed to the case top and bottom with a 3/16-in. gap between each board. Splines were slid into the grooves from the bottom edge of the case until they butted into the rabbet at the top of the case. These splines float freely in the grooves with enough leeway to allow for seasonal expansion. A small brad was driven through a drilled hole in the center of each spline at the carcase base to keep the splines in place.

Legs support the frame-and-panel doors

Two frame-and-panel doors were made and fitted flush with the front of the case (see *FWW* #107, p. 67). I added a drop-down leg to the back of each door's lock stile to support the weight of the opened doors and the heavy tools stored on them, as shown in the photo on p. 68. The legs are short enough to fold up and stow out of the way so that the doors close freely, as shown in the photo at left. A lock and strike plate were mortised into the lock stiles and a keyhole cut in the face of one door. I had planned to add wooden door knobs to the completed case, but the two plain doors were so striking without them that I never added the knobs. I use the key in the keyhole to open and close the case.

Base puts chest at convenient height

A four-legged base, 28 in. high, was made to hold the chest, so I don't have to bend over to get to my tools. The chest sits on the rails, and cove molding glued to the top edge of the rails hides the joint between the chest and base. The chest is not screwed to the frame; its weight is sufficient to keep it from moving. The legs were tapered on a bandsaw and cleaned up on the jointer. Corner blocks strengthen the base and add support for the tool chest, as shown in the bottom photo.

I have been using this chest for the past six-and-a-half years and am very pleased with it. The only thing I would change is the excessive number of drawer slide grooves in the drawer dividers. I also have considered replacing the stand with a lower case for storing items like routers and drills. But this one looks just too nice to change. □

Steven Bunn is a woodworker in Bowdoinham, Maine.

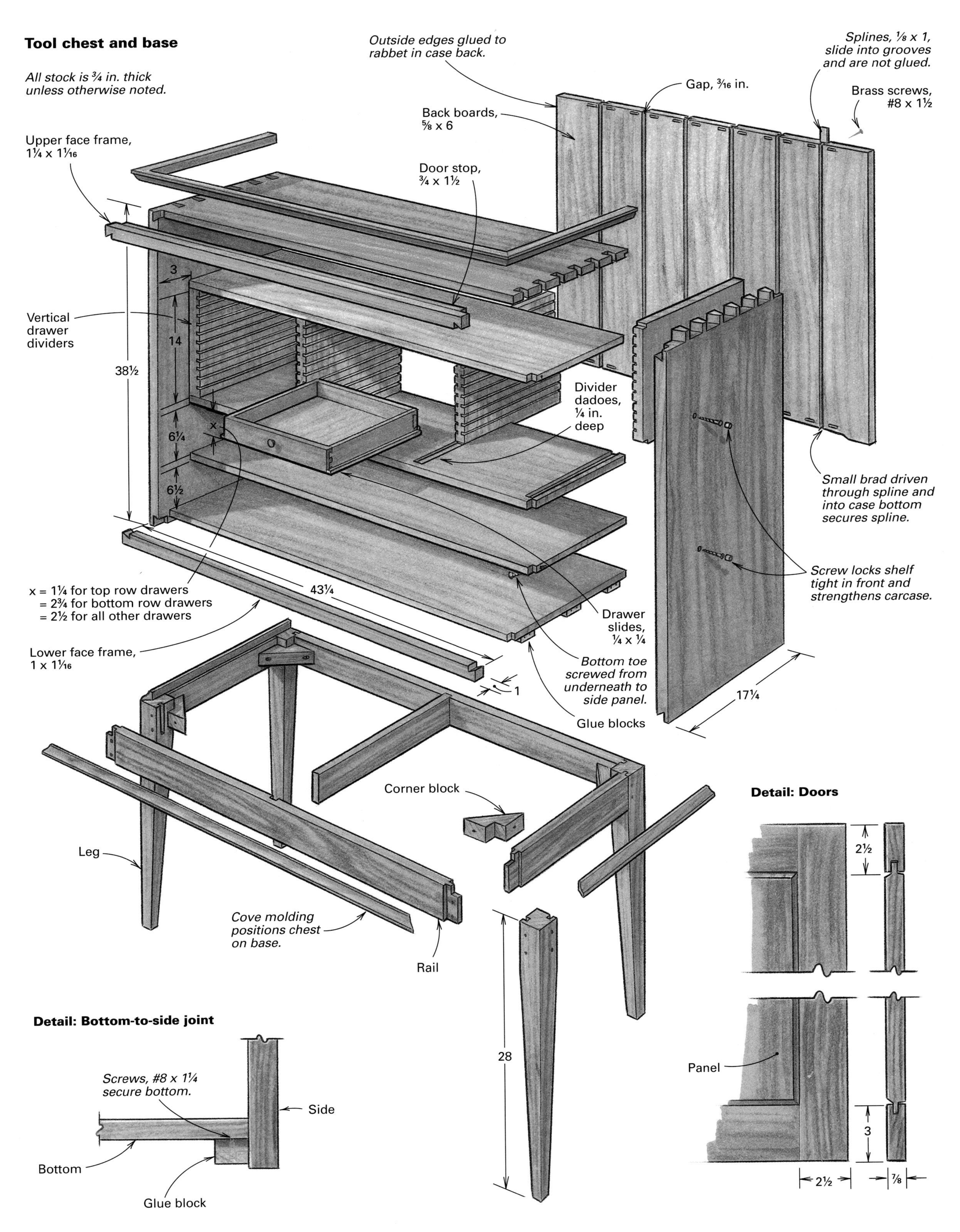

Drawing: Bob La Pointe

Build a Better Sawhorse

by Voicu Marian

Useful shop fixture offers good layout and joinery exercise

I made my first pair of these sawhorses a few years back while remodeling my house because it was uncomfortable working stooped down on the floor. With a hollow core door on top, I had a fairly sturdy workbench that could be moved easily from one room to the next. After finishing up in the house, I took them back to the shop, and that's where they've proven their value.

My workbench always seems to be cluttered with tools. Before I made these horses, I often used the tablesaw as an auxiliary bench. That worked well as long as I didn't need to cut anything.

Now I have a second workbench: A pair of these sawhorses provides a strong, stable base; a couple of thick, heavy planks atop them form a perfectly serviceable benchtop; and a plank across the stretchers makes a good shelf for bench planes and other larger tools that normally clutter a bench surface. I clamp horses and planks together for stability and use C-clamps and bar or pipe clamps in lieu of vises, dogs and bench stops. When I'm finished with the bench, it disassembles and stores easily.

What makes these horses different from most, though, is the joinery. I first saw this half-lap, half-dovetail joint (see the drawing) used by an old carpenter when I was growing up in Romania. It's a strong joint, not too finicky to cut—especially in softwood. The joint gives these horses greater strength and rigidity, a much longer life and, as a bonus, a nice look. Also, the practice you gain in laying out and cutting the joinery in construction lumber will transfer to the fine work you do in hardwoods.

Construction sequence

I dimension all my stock first and then bevel all the edges with a block plane. To ease assembly and ensure consistency, I nail together a quick, simple set-up jig, consisting of three pieces of scrapwood on a plywood base (see the photo).

I determine the angle of the legs by eye rather than by using any mathematical formula. I hold two legs upright and adjust their spread until it looks right. Checking with a protractor for future reference, I read 35°.

I cut the notches at the top of the legs for the saddle first, space the legs with a block the same size as the saddle and then lay out the short end stretchers. I lay out and cut the half-lap first, scribing from the insides and outsides of the legs. I mark out the dovetails on the top side of the stretcher at 8°, cut them and scribe around them with a sharp pencil onto the legs (see the photo). I cut and chisel out the leg to receive the stretcher. When the joint is assembled, leg and stretcher should be flush.

With all four end assemblies complete, I stand up a pair at a time and install the saddle, leaving a 4-in. overhang at each end. This provides a wider support for the boards I use as a benchtop as well as clearance for my feet. For now, one screw holds it together. Next I adjust the sawhorse so it's square to the surface it's standing on. Then I place the long stretcher across the short ones. I center it and mark it for length and for the shoulder of the half-lap.

The rest of the process is the same as for the short stretchers, except I put the dovetails on opposite sides at each end. I do this more for aesthetics than for any structural reason. I glue the long stretcher in place and screw it from below. I then put two more screws on each side of the legs for a total of three screws into the saddle at the top of each leg.

The last thing I do is cut the tips of the feet, so they don't rock. To mark them, I lay the sole of my square flat on its side, scribe around each foot and then saw them off. □

Voicu Marian works wood in Alliance, Ohio.

The well-built sawhorse

Optimal dimensions for these horses depend on the function for which they're intended and on individual height and preference. For someone of average height, 32-in. horses make a good base for an auxiliary workbench, and 24 in. horses are about right for an assembly and finishing platform.

***A simple, nailed-together jig speeds layout** and ensures consistency from horse to horse. Here, the author scribes around the end stretcher dovetail to cut out its mortise in the leg.*

From *Fine Woodworking* (March 1994) 105:75

Photo: author; drawing: Maria Meleschnig

Versatile Shop Storage Solutions

Wheels and wall cleats make for easy rearranging

by Joseph Beals

During the 10 years I worked in a cellar shop, I installed cabinets, drawers and open shelving wherever room allowed. The results were typical: I knew where to find everything, but there was little order to the method, and junk and dust were a chronic problem.

When I moved to a converted garage building, I left those built-ins behind. I packed tools, hardware and supplies into dozens of 5-gal. buckets, and I worked out of them for the next year until the new shop was at last functional. To avoid recreating the past, I designed a new storage system that remedies many of the usual irritations. I resolved to minimize any sort of generic storage that invites accretions of dust and junk. This meant little or no open shelving, no big drawers under the bench and no casual boxes or bins.

Finally, with the agony of moving so close behind me, I wanted a fully portable storage system. And I wanted a system that could be moved around easily.

***These movable cabinets** keep tools stored, on slide out shelves or in drawers, neat and dust-free. Casters make the base cabinets mobile while a cleat-mounting system allows the wall cabinets to be easily rearranged.*

Mobile cabinets

With these goals in mind, I built a set of floor and wall cabinets, as shown in the photo above, which offer exceptional utility in concert with a pleasing, traditional appearance; I also built special wall storage racks, as discussed in the box on p. 76. The floor cabinets are mounted on casters and incorporate a series of guide rails for shelves or drawers. The wall cabinets hang from simple wall-mounted cleats (see the photo at left on p. 77) and include integral dadoes to allow any combination of plain or purpose-built shelving. To cut costs, I built the cabinets from a variety of wood species using leftover stock and cutoffs, including quartersawn white oak, black walnut, mahogany, elm and cherry. All cabinets include paneled doors, ½-in.-thick birch plywood backs and straightforward joinery.

Building considerations

Many woodworkers rely upon detailed, measured drawings as the final design stage, but that's like committing a melody to manuscript without opening the piano for a trial run. Unless you have an extraordinary ability to assess light and shadow, mass, proportion and function on paper, you risk building a sterile, technocratic piece. Remember that a final measured drawing merely records the component dimensions of a functional, aesthetically pleasing prototype.

I used molded frames, raised panels and polished finish to create a display for clients visiting my shop, but there are many simpler options. Pine frames made on the tablesaw, router table or entirely by hand, together with ¼-in.-thick plywood panels and a paint finish are attractive and require no special tooling. A solid, flat panel, rabbeted around the perimeter to fit the frame grooves, is fully traditional and easy to make. If you are new to frame-and-panel work, these alternatives are a practical and satisfying introduction.

For a contemporary appearance, substitute plywood for frame-and-panel construction. Plywood cabinet sides can be grooved to house shelving or drawers, eliminating the guide rails required by a frame-and-panel carcase. Plywood has some drawbacks, however. A-C fir plywood is generally too crude for cabinetry, but ¾-in.-thick birch plywood, which is the least expensive alternative, will cost about $45 per sheet and is best suited for a paint finish. Also, exposed plywood edges must be banded for a good appearance, even under paint. Commercial banding veneers with a hot-melt adhesive are easy to apply, but shopmade solid edge-banding is more robust, and it looks better.

Photos: Charley Robinson

Fig. 1: Base storage cabinet

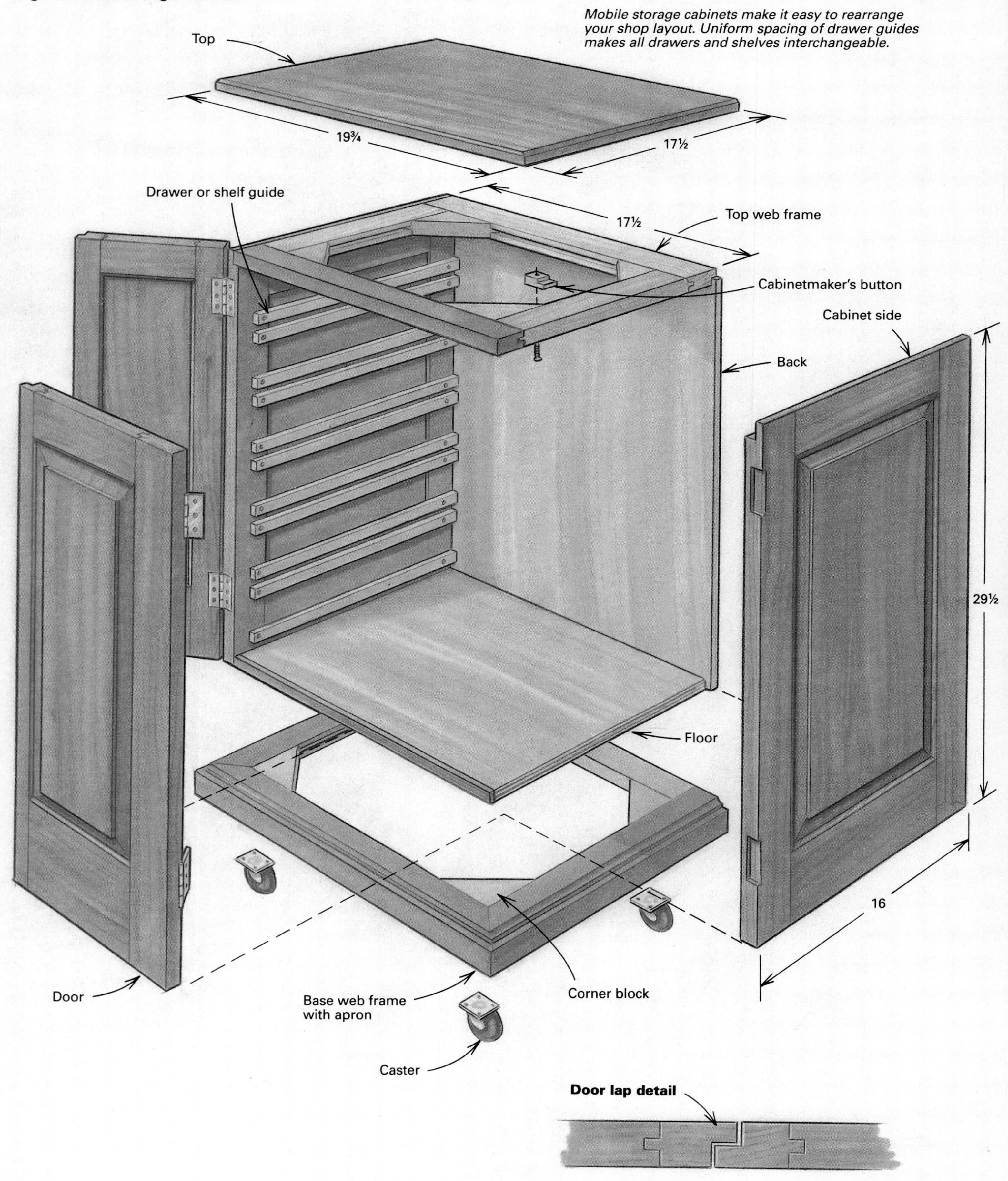

Drawings: Bob La Pointe

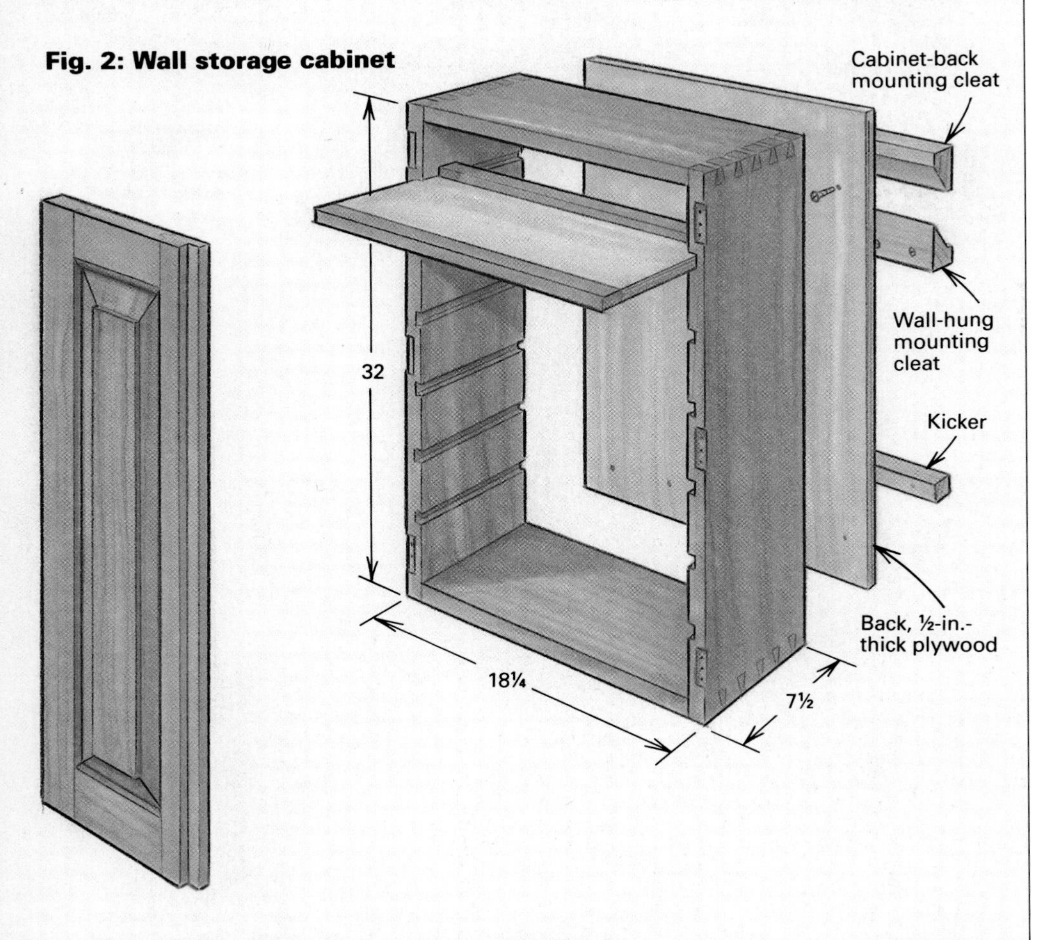

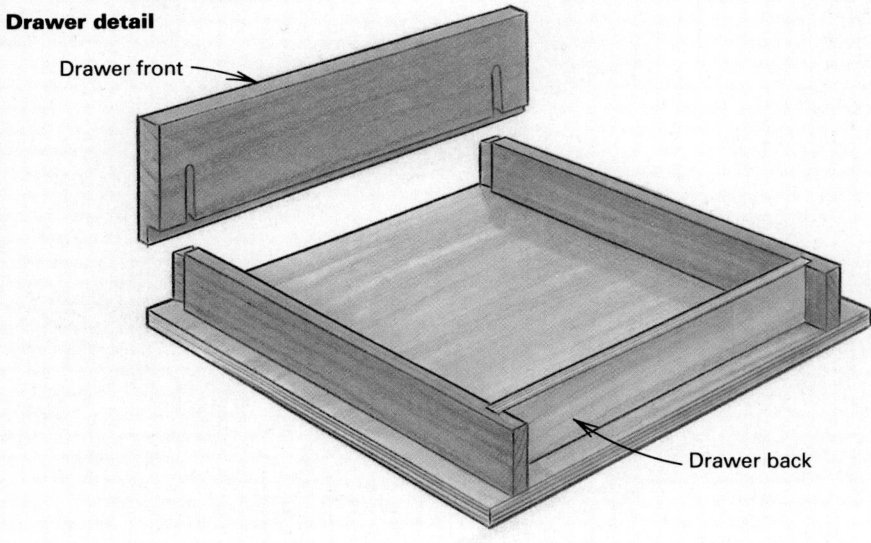

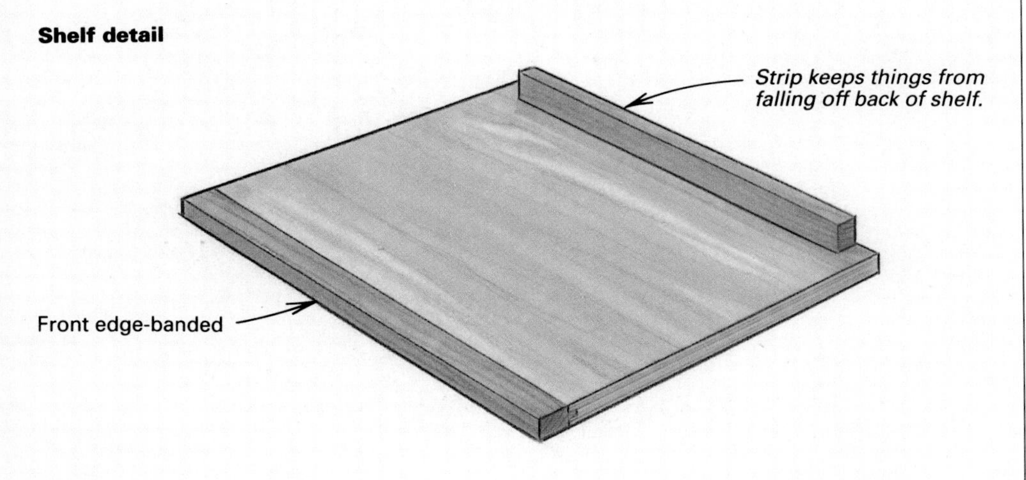

Base cabinets

I built all the cabinets in multiples for maximum benefit of bench and machine time, but I'll describe the construction as if I were making only one of each, starting with a floor cabinet. I began with the frame-and-panel sides (see figure 1). The stiles are equal in length to the height of the frame, but stile and rail widths and the length of the rails, are determined according to personal preference and the method of joinery. Mortise-and-tenon joinery, for example, requires additional length on the rails for the tenons.

I used a matched set of cope-and-pattern cutters on the shaper to machine the frame, but there are many other equally suitable methods (see *FWW* #86, pp. 76-79). To ensure accuracy, I took panel dimensions off a dry-assembled frame. After preparing the stock, I wasted the bulk of the bevel on the tablesaw, then finished fielding the panel with a panel-raising cutter on the shaper. (For more on machining panels, see *FWW* #94, pp. 65-69.)

When the two sides were assembled and cleaned up, I used my shaper to cut a rabbet on the inside back edge to receive the plywood back and another across the inside top edge to house the upper web frame. Finally, I installed the maple drawer and shelf guides. The guide spacing is uniform, so any drawer will fit any space. And drawers and shelves are interchangeable, as shown in the bottom right photo on p. 77. To make this job accurate and quick, the guides are prepared in advance with counterbored screw holes and positioned with a series of spacers, as shown in the top right photo on p. 77. That ensures consistent, square placement. I load the screws into their holes, and run them in with production-line speed.

An upper web frame keeps the top of the carcase square, provides fastening for the solid top and an upper stop for the doors. Notice that the front member of the web is full length and is the only part that need be primary wood. For a run of several cabinets, using secondary wood for the sides, back and corner blocks can save an appreciable amount of stock. A lower web frame is the load-bearing part of the cabinet base. The lower web is exposed on the front and both sides with the two front joints mitered for a better appearance.

Both web frames are of traditional construction, as shown in figure 1. I used a shaper to machine the full-length grooves, and I cut the tongues on the tablesaw. I also splined the lower web miter joints for strength and convenience of positioning during assembly. To avoid juggling eight

From *Fine Woodworking* (March 1994) 105:79-83

parts at once, the web frames are first glued without corner blocks. When the glue has set, the corner blocks are installed in a second operation.

A skirt on the front and sides of the lower web frame gives visual mass to the cabinet base and shrouds the casters. After the skirt is applied, I used a pair of cutters on the shaper to machine a simple reverse ogee (cyma reversa) molding detail on the front and sides, giving a graceful transition from the carcase to the base. Finally, a plywood shelf is affixed to the top of the lower web frame, with the front edge-banded with the cabinet's primary wood. The shelf provides a cabinet floor and positive positioning for the cabinet sides, and the front edge serves as a lower door stop.

All the cabinet components are screwed together, so assembling the cabinet is quick and easy. The two sides are fastened to the lower web frame with screws driven up from below, just inside the skirt. The upper web frame is screwed down into the sides from the top and the plywood back is screwed into the rabbets that house it. I used no glue in the assembly, which makes it possible to take the carcase apart for any reason. I used shop-grade elm for a serviceable top on all floor cabinets. The tops are given a half-round profile on the front and sides, and they're fastened to the upper web frame with traditional buttons (see figure 1 on p. 74).

The cabinet shelves can be solid stock or sheet goods, as preference dictates. I used

Wall racks for clamps, lumber or shelves

With tools and hardware stowed out of sight in the new cabinetry, I was still left with a pile of clamps along one wall and a stack of lumber on the floor. My solution to both these problems was the same, as shown in the photo below and the drawing at right. The lumber rack is of identical construction to the pipe-clamp rack but built with more substantial members.

The racks are easily built by sandwiching spacer blocks between two vertical pieces to create mortises that hold the support arms. The support arms are angled on their lower edges, and the wedges that hold the arms in place have a matching angle on their upper edges. To secure the support arm, slip it into the mortise, push the wedge into the mortise below the arm and tap the wedge into place with a hammer. The arms can hold clamps, lumber or even shelves for storing other small items.—*J.B.*

Wall racks for clamps *and lumber storage are easily made by sandwiching spacer blocks between a pair of vertical supports. Support arms slide into mortises and are secured with a wedge.*

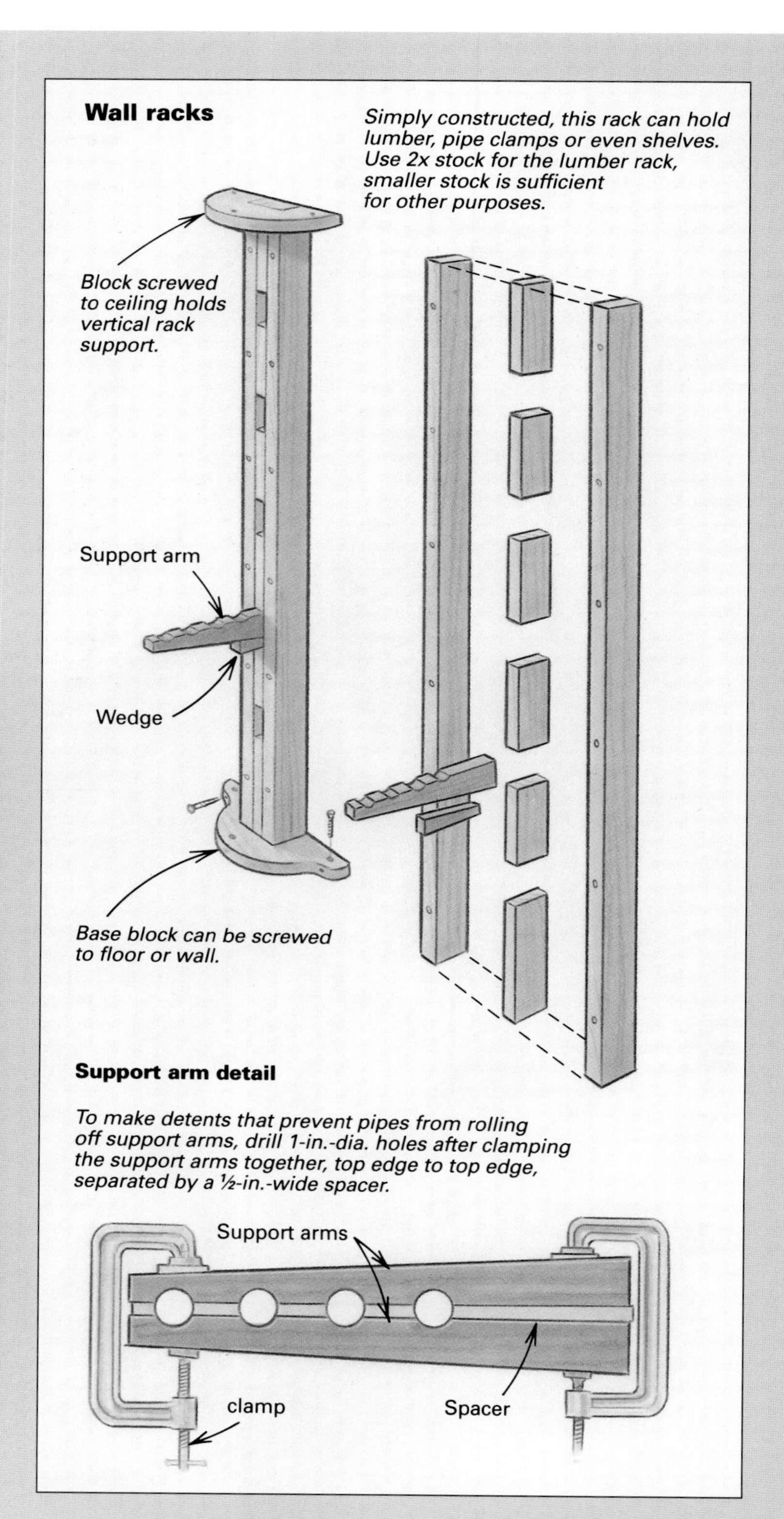

Cleats allow cabinets to be moved—*Strips ripped at 45°, one screwed to the wall and one to the cabinet back, (above) make it easy to rearrange cabinets. A kicker screwed to the cabinet back near the bottom makes the cabinet hang plumb.*

***Installing drawer guides** (above) is done with production-line speed by drilling counterbored screw holes and positioning the guides with spacers.*

***Drawers and shelving are interchangeable** because the guide spacing is uniform (right). Salvaged stock was used for the veneered drawer fronts.*

½-in.-thick birch plywood, banded on the front to match the cabinet wood. The shelves pull out easily on the guides, and thin cleats glued to the back keep things from falling off the back edge.

To keep the design simple, I built the drawers as a box fastened to a shelf, as shown in the detail on p. 75. The two sides engage the front with sliding dovetail joints, and the front of the shelf fits in a rabbet on the drawer front. I screwed the sides and back to the bottom from below. The cabinet will hold six shallow drawers, but deeper drawers can be made by doubling or tripling the spacing module.

Building wall cabinets

Construction of the wall cabinets is simplicity itself (see figure 2 on p. 75). The solid carcase can be assembled in a variety of ways (see *FWW* #104, p. 75), but through-dovetails offer the strongest and best-looking joint. I cut all dovetails by hand, which took less than an hour for each cabinet. Before the carcase was assembled, I used a dado set on the radial-arm saw to cut six shelf dadoes in each side. Using steel shelf standards or a series of holes for shelf support-pins would provide a greater range of spacing options and a cleaner look, but my prior experience with these methods was unsatisfactory. A fully housed shelf never tips out nor does it require a store of mounting pins or brackets, which are typically missing at the moment of need.

Finally, I machined a rabbet on the inside back edges to house the plywood back. The rabbet is full length on top and bottom and is stopped a ¼ in. short of the ends on each side. I cut the rabbets on the shaper and cleaned the stopped ends with a chisel after the carcase was assembled. The back rabbet could also be routed with a bearing-guided bit or on a tablesaw. When the back is in place, a hanging cleat is fastened near the top (see the photo at left above), and a kicker is fitted near the bottom to keep the hung cabinet vertical.

Frame-and-panel doors

All cabinets are provided with a pair of narrow, paneled doors. A single-wide door might seem simpler, but the sweep can be awkward, especially on a floor cabinet in a restricted space. The doors are constructed like the floor cabinet sides and for appearance, have the same dimensional proportions of stiles and rails. Notice, however, that the two inside stiles are half width and give the appearance of a single, full-width stile when the doors are closed. The stiles are also half-lapped, as shown in the door-lap detail on p. 74, such that a single catch on the right-hand door keeps both doors closed. I cut the inside stiles wide to assemble the doors and machined the lap joint after assembly.

I used aniline dye to stain all but the cherry cabinet. Because cherry darkens so rapidly and dramatically, it is generally better not to color the wood under the finish. All cabinets received a half dozen coats of shellac, and after the last coat is well rubbed out, I applied a beeswax polish for a soft, lustrous finish. □

Joseph Beals is a builder and custom woodworker who lives in Marshfield, Mass.

Power-Tool Workbench

Tool storage within an arm's length of the job

by Lars Mikkelsen

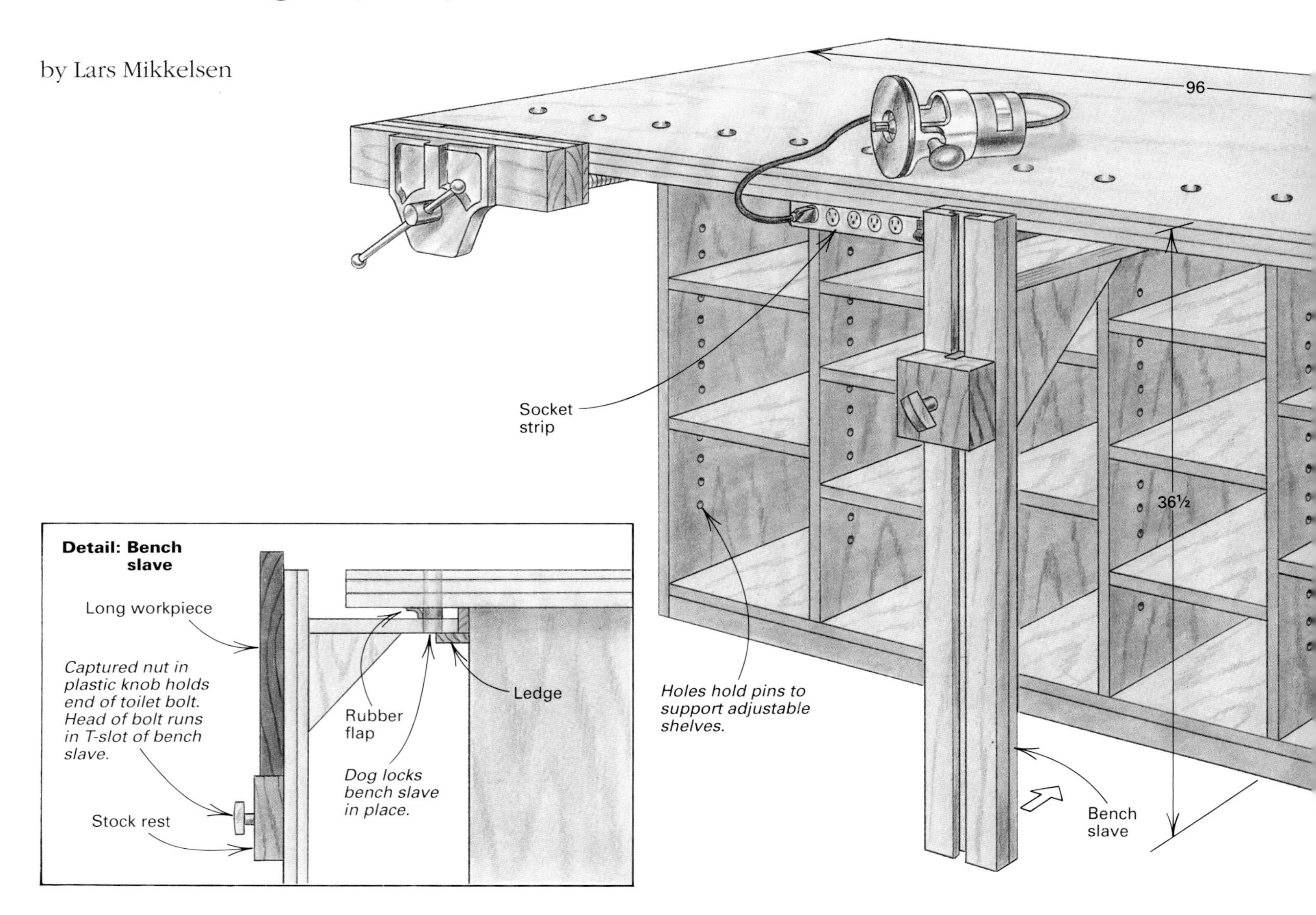

Space is at a premium in my small shop, so the more functions any one thing can serve the better. I had two things that needed improvement—my hand power tools were cramped in a small cabinet, their cords always entwined, and my bench needed a good base. So I decided to kill two birds with one stone and build a base cabinet for the bench with cubbies for my tools.

These cubbies have worked out very well for me. Each tool has its place, where I also keep the miscellaneous wrenches and screwdrivers needed for that particular tool. The small size of the cubbies makes the tools much easier to find than if they were stored on long shelves. The cords never get tangled, and it's so easy to get and put away a tool that I avoid the usual clutter on the benchtop. The power strip that I attached to the bench makes it possible for a tool to be in its cubby while still plugged in ready to go.

Bench slave holds long stock*—The author made a bench slave with a brace at the top that locks into the 1-in. dowel bench dogs he uses. Round dogs are easier to make and install than traditional square dogs.*

From *Fine Woodworking* (July 1993) 101:76-77

Photo: Sandor Nagyszalanczy

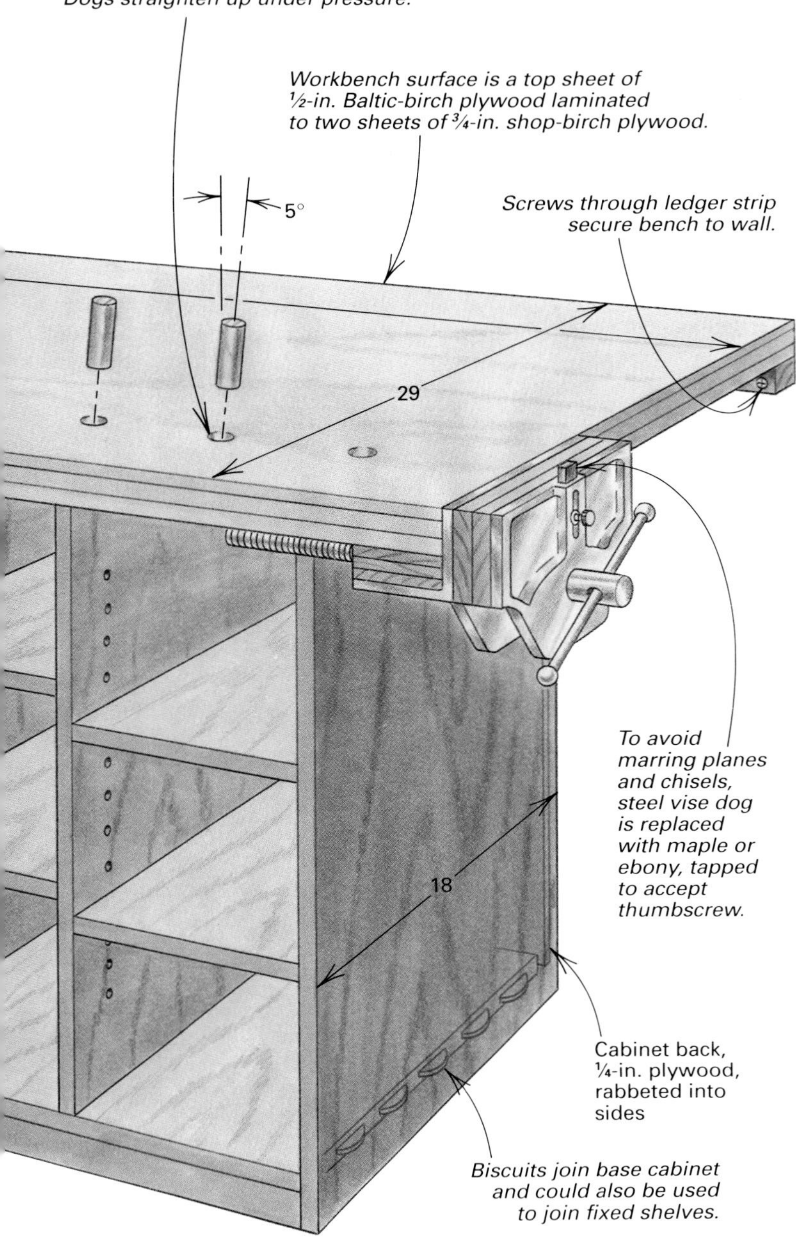

Quick-access tool cubbies—I made the base cabinet from ¾-in. birch plywood edged with ¼-in. strips of solid birch and biscuited together, as shown in the drawing. The biscuits could be replaced with tongue-and-groove joints or dadoes and rabbets, but biscuits are the simplest. I measured my biggest tool to determine the maximum width and depth of the sections. The desired final height of the benchtop sets the base height, and the shelves are adjustable. The dimensions can all be adapted to your own situation, but it is helpful to keep the combined width of the benchtop (mine is 29 in.) and depth of the base cabinet (mine is 18 in.) below the standard 48-in. width of a sheet of plywood, so you can use the cutoffs from ripping the top to make parts for the base. And while I used all the space for cubbies, one of the sections could easily be set up to hold simple sliding shelves for bit storage. The shelves could slide in dadoes cut across the width of facing vertical dividers before assembly.

I used ¼-in. plywood for the back of the base and anchored my bench to the wall with 3-in. screws driven through a ledger strip on the underside of the top. For a freestanding bench, I would recommend, at minimum, a ¾-in. back and a hefty face frame to add stiffness against racking. For maximum strength in a freestanding bench, cubbies could be made to fit beneath a traditional mortise-and-tenon trestle base.

A plywood work surface—The top of my bench is made from two layers of ¾-in. shop-birch plywood and one layer of ½-in. Baltic-birch plywood. Unlike shop-birch plywood, which has a core of thick softwood veneers between thin outer layers of birch, Baltic birch is all birch with a core of thin, high-quality veneers, free of voids. (Baltic birch sheets are often sized metrically and will run approximately twice the cost of shop birch.) This sandwich of shop birch and Baltic birch makes the benchtop amply stiff, and the Baltic birch has a surface hard enough and thick enough to withstand some abuse. I laminated the three sheets of plywood with Liquid Nails construction adhesive. I did not have any way of clamping something this big, so I used lots of screws coming up from the bottom. I removed the screws once the adhesive had set, so I wouldn't run into them later when drilling for the bench dogs or other fixtures.

A new twist on old dogs—I mounted two Record #52½ ED vises to the top, one as an end vise, the other as a front vise. Both have wooden jaw faces. I tapped the metal jaws, so I could change the wooden faces easily without removing the vises from the bench. The front vise has oversized jaws to get a better grip on large pieces. To make bench dogs, I cut up a 1-in.-dia. dowel. I drilled a series of 1-in. holes for the dogs in line with the end vise dog. The holes angle toward the vise at 5° so the bench dogs straighten up under pressure. To keep the dogs from sliding down when in use, I tacked small strips of rubber to the underside of the bench, partially overlapping the dog holes. But I was afraid vigorous pounding on the bench might make the dogs fall out, so I screwed and glued a ledge to the base that supports the inside half of the dogs. I can easily reach under the bench to push the dogs up, and when not in use, they are firm against the ledge.

Long stock support—For the times when I have a long piece of stock clamped in the front vise, I made a bench slave to support the free end (see the photo). The outer face of the slave leg is in the same plane as the inner jaw of the shoulder vise. I use the bench dogs and the ledge beneath them as a way of locking the slave to the table. Instead of making it freestanding, with feet that might get in my way, I built a kind of peg leg with a brace near the top that slides under the benchtop and rests on the ledge beneath the bench dogs. I drilled a slightly oversized hole through the brace, so it can easily be locked in place under any of the bench dogs.

The stock rest, a block of solid wood, is attached to the leg with a toilet bolt that slides in a T-slot (as shown in the drawing detail) and can be locked at any height on the slave. To make the leg, I cut a shallow groove in a piece of ¾-in. solid wood and glued it, grooved side in, to a piece of ½-in. plywood; then I cut a narrower groove in the outside face, forming a T-slot for the head of the toilet bolt. A spline glued into the back of the stock rest rides in the stem of the slot.

My bench was relatively inexpensive to build and serves my purpose well. I like the big top, and the vises can hold everything I work on, from big doors to the occasional miniature. Doors on a base like this might look good, but the ease of access would be lost, and in a shop, efficiency comes before aesthetics. □

Lars Mikkelsen is a professional woodworker in Santa Margarita, Calif.

Drawing: Mario Ferro

Flip-Stop Fence for a Radial-Arm Saw

The track-mounted stop is always handy

by Art Duser

"You can't buy any more tools," my wife told me in no uncertain terms, and she added, "the accessories cost too much!" Though she wasn't entirely serious, her wake-up call helped me to rank my woodworking tool needs; I began to think seriously about what I could do without and what I could build myself. Since then I've built a number of power-tool accessories, including the radial-arm saw fence with flip stop that's discussed in this article. Building my own accessories not only saves me money (allowing me to buy more lumber and essential tools) but also provides real satisfaction, borne of self-reliance and successful problem solving.

My fence consists of two 5 ft. lengths of mahogany—a main fence and an extension fence—that I spliced together, as shown in the drawing. A dado in the top of both pieces of mahogany houses a piece of extruded aluminum shelf standard (manufactured by Dorfile, a division of Newell Home Hardware Co., 4533 Old Lamar, Memphis, Tenn. 38118; 901-365-0479). The lips of the shelf standard capture a toilet-flange bolt that extends through a piece of polyethylene (generally available as scrap from a plastics supply house). A knurled thumbnut atop the block permits me to lock it in place. The flip stop is screwed into a hole that I drilled and tapped on the side of the polyethylene block.

The main fence is pinched between the front and rear portions of the saw's table (in the same manner as the stock fence) and is made more rigid near the blade with a piece of aluminum angle that I screwed to the main fence's downward-extending lip (see the drawing). The extension fence can be connected to the main fence either with a pan-head machine screw and a T-nut set into the extension fence (as I did) or with a wood screw through both members into the table below. Both extension and main fences are carefully aligned and then screwed to the extension table.

***Quick, accurate multiple cuts** are all but assured with a good fence setup. The flip stop is angled up in the front to allow the user to push a piece of stock into the fence for a square cut and then flip the stop back down for the cut to length.*

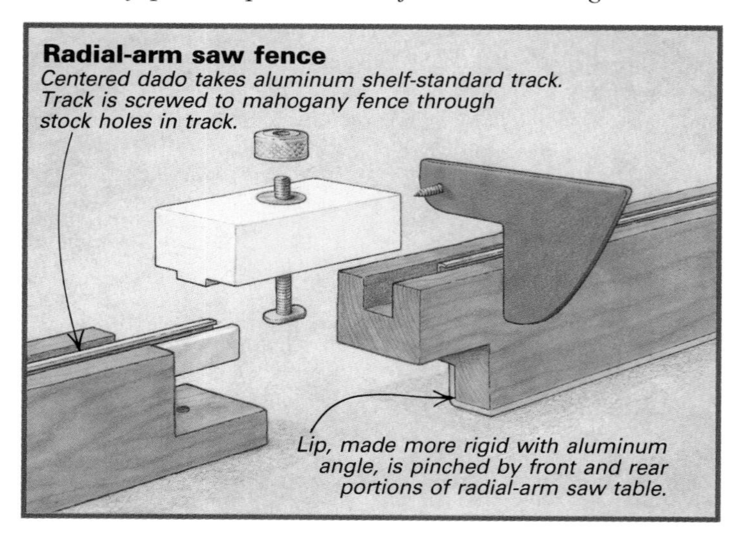

Stop block and flip stop

The first stop block I made was of hardwood and required a mortise on its back side to capture the locking nut for the flip stop's pivot screw. Since then I've made a few of these fences for friends, and I've used polyethylene for all my stop blocks. The polyethylene permits drilling and tapping of the hole for the pivot screw and will hold the screw snug with just enough play.

I've used knurled thumb nuts on the toilet-flange bolt for all my stop blocks (though I'm sure a wing nut would do the job as well), and I embed a flat washer in the top of the block to prevent the thumb nut from wearing into the polyethylene. To minimize friction in the track when the stop block is moved, I file the flat sides of the flange bolt's head slightly. I also set a rubber tack bumper into the main fence near the blade to discourage the stop-block assembly from getting too close to the blade.

I've used both aluminum and phenolic-resin board for flip stops, but plywood would probably work as well. The front edge of the stop curves upward so that it will flip back when a long piece of stock is pushed into it. This makes it easy to trim one end of a board before flipping it end for end to cut it to length.

I use my radial-arm saw almost exclusively for 90° crosscutting, so I've tuned it up to cut dead on. When I need to cut a 45° miter or any other angle, I position the piece I'm cutting with one of a number of precut plywood angles I've made for this purpose. I find this method to be more accurate and a lot quicker than having to reset my saw every time I want a different angle. If you cut angles other than 90° on your saw, you'll want to keep the aluminum back farther from the blade, and you may have to use a block against the flip stop if you are cutting fairly short pieces.

I've only made these fences for radial-arm saws, but I'm sure some ingenious woodworker could modify the design to accommodate a miter saw or even a sliding compound-miter saw. □

Art Duser is a woodworker and a member of The Triangle Area Woodworkers' Club, which meets in Raleigh, N.C. He's also a data communications specialist.

From *Fine Woodworking* (May 1992) 94:49
Photo: Ed Walker; drawing: Heather Lambert

A compression chuck holds a bowl securely at its rim while its foot is turned.

Compression Chuck for a Lathe

Shop-built chuck holds bowls tightly, so you can turn a foot

by Dale Ross

A nicely finished foot on the bottom of a turned bowl is one feature separating the work of a pro from that of a beginner. A well-proportioned foot lifts the bowl and gives it a classic look typical of pottery. Turning a foot also eliminates the mounting screw holes on the bottom of the bowl.

The biggest problem with creating a foot or finished bottom is not how to shape it, but how to hold the bowl in the lathe. Turning the foot is the last thing I do, so the outside and inside waste of the bowl has already been cut away and sanded, leaving no place for mounting screws. That's where my shop-built compression chuck comes in, making it possible to remount the bowl and complete the foot. The real advantage of this system is that once the chucks are made, they can be used over and over again. My set of four chucks will handle bowls ranging from 4 in. to 14 in. dia. The chucks are easy to make and inexpensive, too, because they're made from plywood and mahogany or poplar scraps.

How a compression chuck works

A compression chuck consists of a flexible jaw plate pressed to a curved baseplate by a platen, as shown in the drawing on p. 83 and the bottom photo on this page. A handwheel is screwed to the outboard end of a threaded rod that passes through the lathe's headstock. Tightening the handwheel draws the platen toward the headstock and squeezes the jaw plate between the baseplate and the platen, constricting the jaws of the chuck. As the jaws close in, they grab and hold the rim of a bowl.

The parts are simple. *A compression chuck consists of a platen, jaw plate, baseplate and handwheel (from the left), all connected with a threaded rod. Tightening the handwheel flexes the jaw plate, so it grips the edges of a bowl.*

The jaw plate has a series of evenly stepped ridges to accommodate bowls of varying diameters. The compression chuck shown in the top photo on this page is 11 in. dia. and will accommodate bowls from about 9 in. dia. to 10⅝ in. dia.

The baseplate is turned from plywood. *A template on a lathe bed helps the author shape a camber in the baseplate (colored yellow in the drawing on the facing page).*

A series of steps in the jaw plate *of the compression chuck accommodate a range of bowl sizes (the jaw plate is red in the drawing on the facing page).*

The platen is turned with a crown *to match the dish in the baseplate. The curved platen (green in the drawing on the facing page) flexes the jaw plate.*

Making the baseplate

The baseplate is two pieces of plywood glued together, turned and hollowed out, as shown in the drawing. For an 11-in.-dia. chuck, glue and screw together two pieces of ¾-in.-thick by 12-in.-sq. plywood. Once the glue dries, remove the screws, mark the center and bandsaw the plywood to as large a disc as possible. Temporarily mount the disc to a faceplate, and turn the outside edge true. Then cut a mortise into what will become the back side of the baseplate to match your faceplate (I used a 6-in.-dia. faceplate).

Better yet, leave a faceplate on each chuck. I make extra faceplates from 1-in.-thick aluminum plate, bandsawn round, drilled and threaded to my lathe shaft size. After screwing the aluminum faceplate onto the lathe shaft, I true it round and flat with high-speed steel tools.

To finish up the baseplate for the chuck, remove it from the lathe, and remount it on a faceplate screwed into the turned mortise. On the face of the baseplate, cut a shoulder, and then dish out the face of the plywood, as shown in the top photo. Go about ⅝ in. deep, taking care not to hit the mounting screws. Try to achieve a nice, fair camber. Finally, drill a ½-in.-dia. hole through the center of the baseplate for the mounting bolt and threaded rod.

To help get the shape right, bandsaw a curved template out of ¼-in.-thick plywood (I use a set of trammel points). The offcut will be the template for turning the platen, so hang on to it.

Turning the jaw plate

The jaw plate is the part that actually does the gripping. It's made of two pieces: a thin, flexible plywood backing and an outer ring of solid stock turned to form steps that grip the edge of a bowl. Evenly spaced sawkerfs around the perimeter of the jaw plate allow it to flex as it's squeezed between the platen and baseplate.

For the backing, use ⅛-in.-thick Baltic birch for chucks of 11 in. and less in diameter and ¼-in.-thick Baltic birch for larger chucks. For the outer ring, glue up 8/4 poplar or mahogany into a 12-in. square, and bandsaw it round. After flattening the back of the solid disc and drilling a small hole through its center, I glue it to the plywood backing, but only around the perimeter. When I cut the final step in the blocking, the center section will fall away without a lot of unnecessary lathe work.

One thing that helps keep the center section from being glued to the plywood is a V-groove cut into the back of the disc that serves as a glue dam. The V-groove is cut just outside of where the last step will fall, as shown in the top drawing on the facing page. Apply glue only to the solid wood, outside the stop groove, and glue the solid-wood disc to a slightly larger plywood disc.

Drill a small center hole through the plywood using the previously drilled hole through the solid wood as a guide. This will locate the faceplate on the back side of the plywood. Mount the glued up disc on your lathe, and turn the outside diameter to match the inside diameter of the shoulder turned into the baseplate. Now turn the steps to form the jaws into the face of the solid wood, taking care on the last step not to cut into the plywood (see the center photo). Slightly undercut the sides of each step for a better grip. I make the steps the same width as my parting tool (⅜ in.), so I can cut each step quickly and accurately without measuring. The screws from the faceplate hold the unglued center area in place while turning. Once off the lathe, this center area of the hardwood disc should come right out.

Radial sawcuts, ¼ in. wide (cut from the perimeter of the disc to within 3 in. of the center) divide the disc into eighths and allow the jaws to flex during compression (see the bottom photo on p. 81). If the jaws seem too stiff, make the radial cuts a little longer. A ½-in.-dia. center hole provides clearance for the threaded rod.

From *Fine Woodworking* (July 1995) 113:75-77

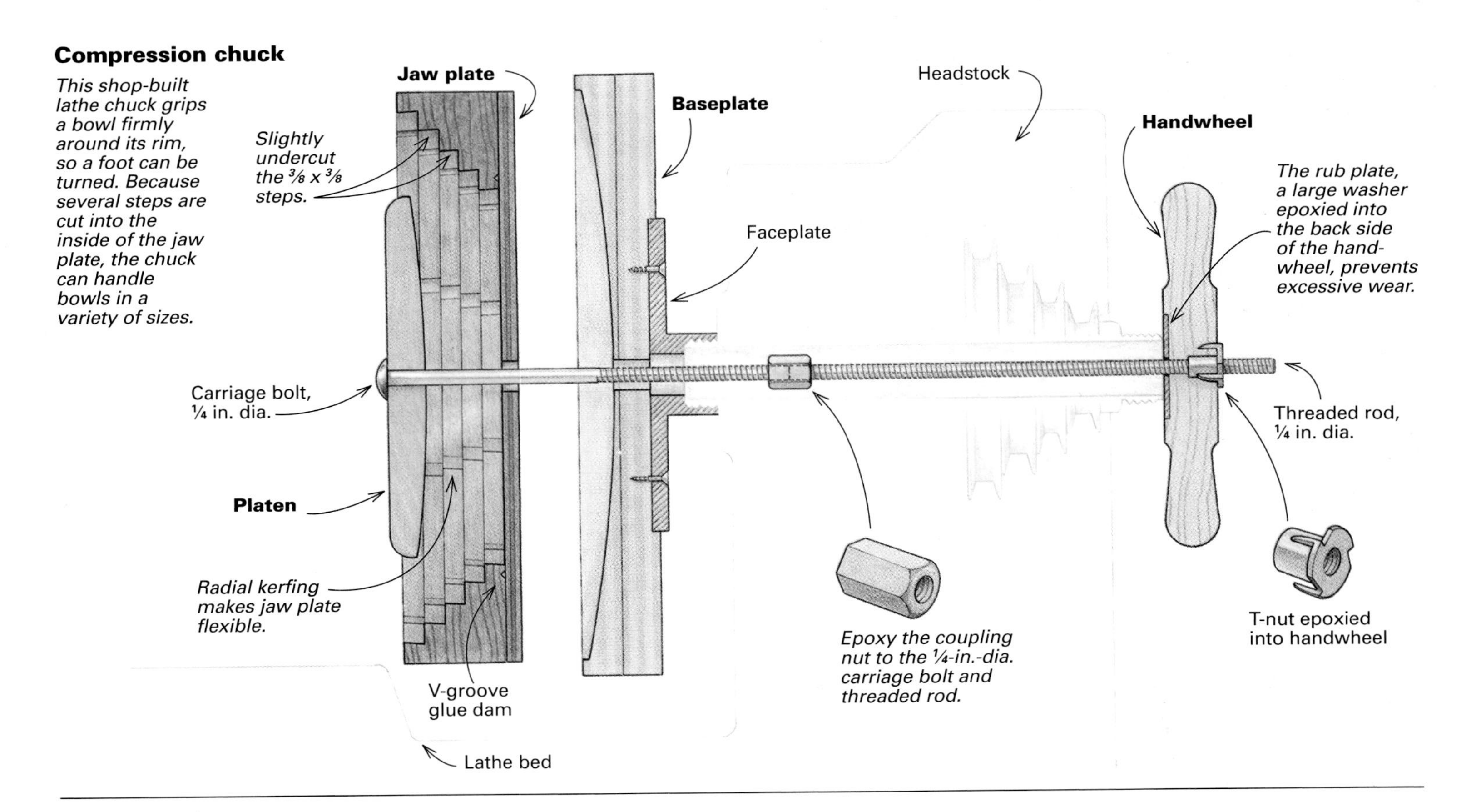

Compression chuck

This shop-built lathe chuck grips a bowl firmly around its rim, so a foot can be turned. Because several steps are cut into the inside of the jaw plate, the chuck can handle bowls in a variety of sizes.

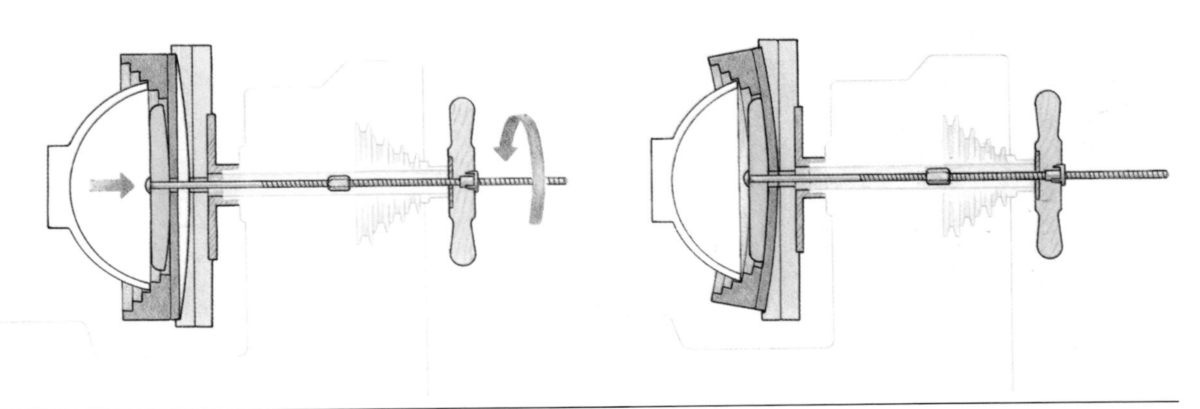

How it works

With the handwheel loosened, the bowl slides easily into the flexible jaw plate. As the handwheel is tightened, the jaw plate is compressed by a curved platen and captures the outer rim of the bowl.

The platen and handwheel

The platen is turned from another piece of 3/4-in.-thick Baltic-birch plywood. Mount a bandsawn, round piece of plywood on the lathe. Turn a crown into the face, matching the camber of the dished-out baseplate. Here's where the other half of the template comes in handy (see the bottom photo on the facing page).

Drill a hole, and insert a 1/4-in.-dia. carriage bolt from the flat side of the platen. Attach a length of 1/4-in.-dia. threaded rod to the end of the bolt with a coupling nut, and then epoxy the joint. The bolt/rod combination should be long enough to pass through the platen, jaw plate, baseplate, lathe headstock and handwheel, as shown in the drawing.

The handwheel, which tightens the jaw plate around a bowl, is turned from hardwood. Epoxy a large washer to the inside face of the handwheel to act as a rub plate. This washer must have an inside-diameter hole large enough to allow the threaded rod to pass through it and an outside diameter large enough to cover the end of the lathe's spindle. Insert a T-nut into the outside face of the handwheel so that it can screw onto the threaded rod. Put the whole rig together on the lathe, and then hacksaw off any extra threaded rod. Leave enough of the threaded rod to engage the nut in the handwheel completely when the jaws are fully relaxed.

Using the compression chuck

With the chuck mounted on the lathe and the lathe's spindle locked, hold the bowl into the closest-fitting step of the chuck. For in-between sizes, I tape small pieces of 1/8-in.-thick plywood to each jaw of the next larger step with double-faced tape, but this is rarely necessary. Tighten the handwheel securely while holding the bowl solidly to the bottom of the step.

The closer a bowl's shape gets to perpendicular at the rim, the less secure the bowl is in the chuck. In this situation, I bring the tailstock up and sandwich the piece in with a long, blunt insert in the revolving center, allowing room for the tool-rest base, as shown in the top photo on p. 81. A center cone, which needs to be cleaned up with a sharp chisel, remains after turning. With light cuts and moderate spindle speed, I can turn a foot on a variety of bowl sizes without any problems. □

Dale Ross is a professional turner in North Yarmouth, Maine.

Drawings: Heather Lambert

Make an End-Boring Jig

Adjustable drill-press setup simplifies difficult drilling jobs

by Jeff Greef

***Boring the ends of long workpieces** used to present problems for the author, Jeff Greef, until he devised this end-boring jig for his drill press. A pair of platens mounted to the press's table allows adjustments in and out and left and right.*

I've built a fair number of custom doors and windows in which I've joined the stiles to rails with dowels. Until recently, I relied on doweling fixtures to position holes. Although fixtures are quick to use, I found them lacking accuracy, particularly for large dowels. The problem is not in locating the fixture's bit guide precisely, but rather it is guaranteeing that the bit drills straight and true.

A horizontal boring machine could solve the problem. But while one of these machines is neither hard to use nor difficult to build, it would eat up precious space in my already cramped shop. Besides, it seemed redundant to buy a motor, bearings and a thrust mechanism when I already had all those things standing in the corner in my drill press.

Boring holes with a drill press is very accurate, but how do you bore holes in the ends of long workpieces? If you turn the press table vertically, clamp a fence to it and secure the work, the setup is still quite limited in terms of making fine adjustments. So with adjusting (and readjusting) in mind, I made an end-boring jig that mounts to my drill press, as shown in the photo on this page.

Drill-press mounting logistics

Before you build the jig, you need to figure out how you will mount it. Although each type of drill press may require a slightly different setup, you should be able to adapt the principles I used to mount a jig to your press. First, my jig is designed for floor-model presses. If you have a bench press, you'll have to bolt its base to a workbench with the spindle overhanging the edge. You can extend the jig to the floor as I did (see the drawing). Second, the jig is made for presses with at least 14 in. of swing to get the depth to the column needed for mounting. Third, the jig is built for presses with tables that both tilt and swivel. By swiveling the table's arm 90° and tilting the table vertically, you can bolt or clamp the jig to it. If your table doesn't tilt and swivel, remove it. Then make a wooden outrigger with a yoke to clasp the press's column. Mount the jig to the outrigger in line with the spindle.

Designing and building the jig

The boring jig has two main parts: a fixed platen and a movable platen. The fixed platen bolts onto the drill-press table. The movable platen attaches to the fixed one with hinges on one side and adjustment bolts on the other. The hinges allow the movable platen to be positioned in and out from the press's column. The adjustment bolts fine-tune the alignment to the bit. The bolts work in a push-me-pull-you fashion (see the drawing detail on the facing page): One bolt pushes the movable platen away, the other bolt pulls it toward the fixed platen and a spring takes up slack. The movable platen also slides to move work left or right, and an adjustable stop plate sets the height of the work. A toggle clamp secures the workpiece alongside a fixed, vertical fence.

Fixed platen—Because of the jig's 4½-in. depth, I turned the press's table a full 90° and pivoted it to vertical before I bolted on the fixed platen, which is just a piece of ¾-in. plywood. I made the platen 22 in. wide to span the distance to the spindle. On the front of the platen, I fastened two vertical boards: one 7 in. to the left of the press' spindle, one 7 in. to the right. I mounted a pair of hinges to the right board, and I recessed T-nuts in the left one to receive the adjustment bolts. Two boards on the back of the movable platen mate with the hinges and bolts.

Movable platen—The movable platen consists of three layers. The outer two layers of the movable platen are long enough to double as a support because the jig and workpiece are suspended from the drill-press table. The inner layer is hinged and bolted to the fixed platen. The middle layer has a strip of wood screwed along the top of it, so it hangs off the inner layer. I made the middle layer so that it can slide left and right; that way, it's easy to precisely bore side-by-side holes in the same plane of a piece. The outer layer is the work surface and also acts as a spacer to keep work clear of the middle layer's bolt heads. I used fender washers under the bolt heads to avoid splintering the plywood.

Photo: author; drawing: Lee Hov

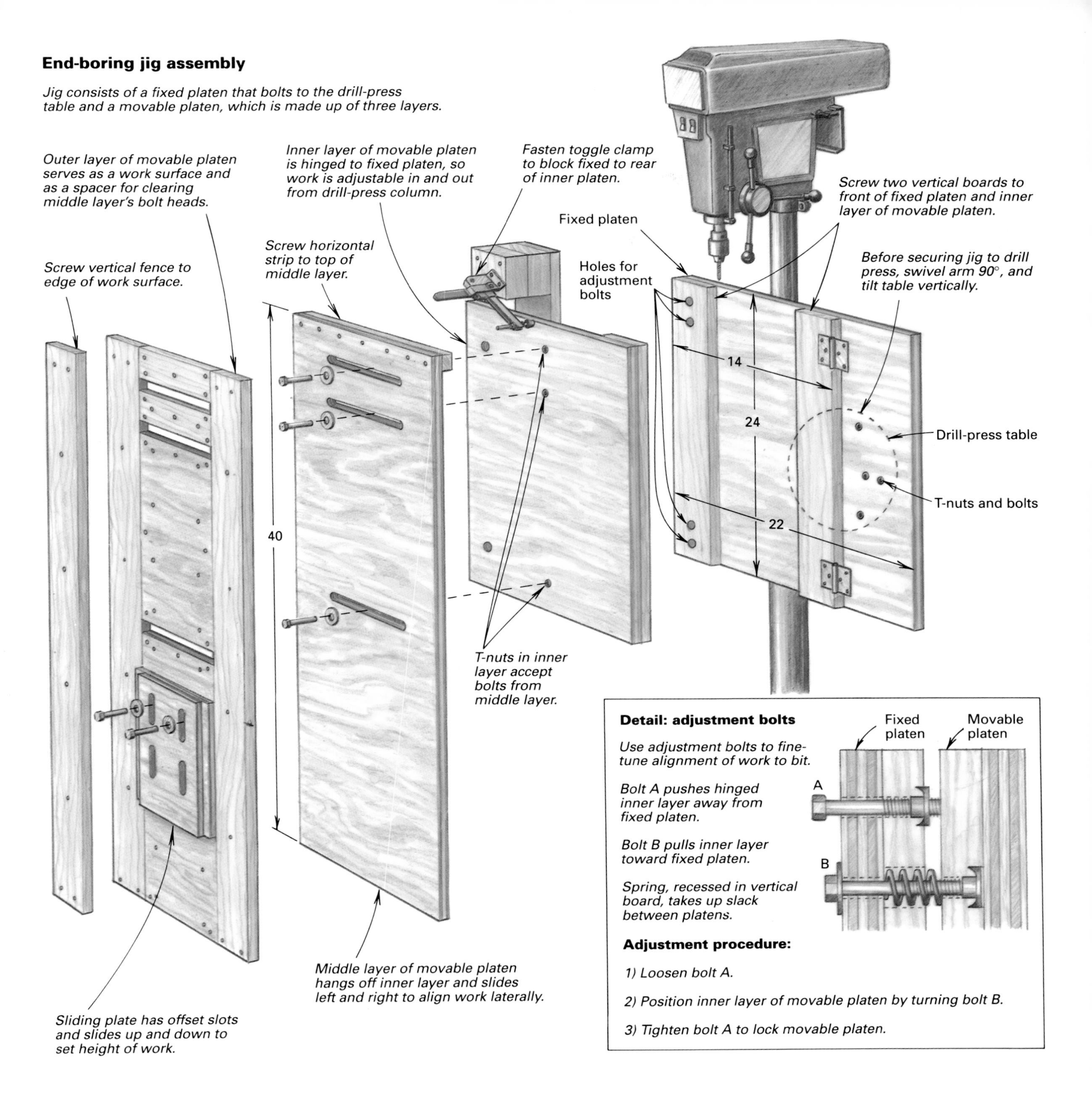

Vertical fence, work clamp and adjustable stop plate—A vertical fence on the left side of the work surface and a toggle clamp fastened to the rear of the inner layer secure the work. The middle layer can still slide without affecting the clamp. During end-boring, the work rests on a stop plate. The plate is slotted to handle pieces up to 40 in. long, and the slots are offset, so the plate can be flipped for other heights. To set the height of a piece, I just slide the stop plate up or down on two bolts.

Using the jig

Once I've set up the jig and positioned the work, I wedge in a pair of shims between the bottom of the movable platen and the top of the press's base to stabilize the jig (see the photo). Even with the platen wedged in place, I can make up to 1/32-in. corrections using the adjustment bolts.

To break in the jig, I bored holes in the ends of rails on 12 interior doors. The end-boring jig proved a real improvement over my conventional doweling fixtures. Once the work was roughly aligned, it was easy to make fine adjustments on test scraps before boring the actual run of holes. I still keep the old doweling fixtures on hand—but only for those rare situations that the end-boring jig won't handle. □

Jeff Greef is a woodworker and journalist in Soquel, Calif.

Scratch stocks—old, new and shopmade—*Whether old like the Stanley #66 (right), new like the Lie-Nielsen #66 reproduction (left) or shopmade (top), these scratch stocks are a simple way to reproduce moldings or create new designs accurately and economically.*

Simple Tools Can Reproduce Most Moldings

Scratch stocks are quick and easy to use and make

by Robert S. Judd

Scratch stocks function beautifully, quickly and economically to duplicate handworked wood trim. By simply grinding or filing a cutter to the appropriate profile, you can reproduce almost any shape molding up to about 1 in. wide. Scratch stocks, or beading tools as they are sometimes called, are readily available new (Lie-Nielsen Toolworks, Inc., Route 1, Warren, Maine 04864; 800-327-2520 or Veritas Tools Inc., 12 East River St., Ogdensburg, N.Y. 13669; 800-667-2986), used (antique tool dealers, garage sales or flea markets) or shopmade (see the photo above). I make mine from a 6-in.-long, L-shaped piece of stock. The cutter fits into a sawkerf, and it is clamped in place with a few screws, as shown in the photo above. The cutters for all of these tools are easily shaped from old scrapers and sawblades or new blanks from Lie-Nielsen or Veritas.

In my repair and restoration business, I often need to duplicate broken or missing moldings. Usually, only a foot or two of the molding is needed: hardly worth the effort of setting up the router and definitely not worth having a cutter ground to match one of the myriad of molding shapes. Besides, no power tool can match the irregularities of the handworked wood found in older pieces.

Scratch stocks and beaders

First made by users as a simple holder for a scraper blade, scratch stocks included a fence arrangement to work a measured distance from an edge. The beading tool was essentially an improved, factory-made scratch stock and included a range of cutters in different sizes and several blanks, custom-filed to fit the user's needs. Adjustable fences for both straight and curved edges were often included. A scratch stock or beader can produce a carbon copy of the original molding by using a cutter that's simply filed to shape.

Photos: Charley Robinson

Filing a cutter to shape*—Almost any profile, up to 1 in. wide, can be filed into blade blanks made from old cabinet scrapers, sawblades or new blank stock (above).*

Matching a molding to a cutter *(below) is crucial to reproducing old moldings. File the cutter to the negative image of the molding. Check the cutter frequently while filing to make sure it is an accurate match.*

Beading is simple with a scratch stock*—Just hold the fence against the stock and make repeated passes (right), about 1/16 in. per pass, until the appropriate depth has been reached.*

Shaping the cutter

To make a basic beaded molding, take a sample piece of beading, a file and a blade blank and set to work filing a negative pattern of the molding, as shown in the top left photo. As you file the pattern into the blade, keep testing its fit (see the bottom right photo). Check the fit frequently because it is fairly easy to file past the desired shape. It's a good idea to leave a 1/8-in.-wide metal strip at either edge of the cutter. Narrower strips tend to bend and lose their effectiveness. Old cabinet scrapers or sawblade sections make good cutters for shopmade scratch stocks. But for my 100-year-old Stanley #66 hand beader, the blanks that Lie-Nielsen makes for his gem-like bronze replicas of the #66 work well. The steel of the new blanks is not hardened, so the blanks are easy to file to shape. After filing them to shape, hone just the cutter's faces in a whetstone to provide a clean cutting edge. I've never found it necessary to harden a cutter once it's filed to shape.

Making moldings

When producing short moldings, I've found it easier to work the edge of my board, as shown in the bottom left photo. For making small beads or moldings, I cut two lengths at once by working both corners of the same board edge. Begin the scraping process by firmly gripping the handles, and push or pull the tool across the board's edge, keeping the handles at 90° to the work. Take small scrapings initially, only 1/16 in. or so at a time. Because stock removal is done by scraping, a small cut gives much more control and does less damage if you slip. As the cutter starts to bottom out, you can continuously adjust the blade so more is exposed. In a surprisingly short time, the molding will start to appear on the edge. If the cutter starts to chatter or jump, you are probably trying to remove too much material, or the grain might be changing; use a little less pressure, or try changing the direction of cut.

One of the handy features of the #66 or the Lie-Nielsen reproduction is the adjustable fence. When cutting two lengths of molding on a board edge, the fence can be set to cut the opposite corner without moving the blade. This lets you produce a surprising amount of molding in a relatively short time. I make several extra moldings, so I can pick the best match to the original.

I like to start the staining and coloring process at this stage because the strips are far easier to handle while they are still attached to a board. Often, I will even do the preliminary finishing and filling at this point for the same reason. It's then a simple matter to trim the finished molding off on the tablesaw. I set the saw fence to leave a little extra material, which I later trim off with a utility knife.

When repairing antique pieces, mark your name and date on the back of the new molding for historical reference. After all, with a matching stain and finish, the repair should be almost invisible.

Other applications

In addition to producing molding patterns, this highly functional family of tools is also effective for routing and inlay work. Because you create the cutters to fit the situation at hand, you are no longer limited to standard router bits.

When using these tools to rout cross-grain, however, it's a good idea to lay out the material to be removed by lightly cutting in the lines with a sharp craft knife. The scored lines help prevent tearout, which could ruin your project. □

Robert Judd is a professional furniture repairer and refinisher in Canton, Mass.

From *Fine Woodworking* (November 1994) 109:86-87

Perfect miters*—Guided by Ed Speas' shooting board (left), a Lie-Nielsen #9 miter-plane easily shaves a 45° miter on molding. The fence is reversible, so the fixture can handle left- and right-hand cuts.*

Fixture doubles as a bench hook *(below). To convert the shooting board to a bench hook for 90° sawing, the author simply removed the miter fence (here resting in the bench trough).*

Shooting Board Aims for Accuracy

Multi-task fixture guides saws and planes for perfect joints

by Ed Speas

Fitting miters has been every woodworker's problem at one time or another. Whether you are making a picture frame or joining molding, if your angle of cut or your piece lengths are not perfect, you have to repeatedly shave a smidgen to get a tight joint. Although a chopsaw or a tablesaw can save time and effort, it may not be the best choice for extremely clean and accurate cuts. If you use a handsaw, it tends to wander if not precisely guided. And even then, I don't know too many folks who can really get consistent forty-fives with a hand miter box alone. Trimming 90° cuts can also be a problem. A sawblade, hand or power, rarely leaves a smooth enough surface. If you sand the end grain, again, you risk introducing error.

You can eliminate these difficulties by using a simple fixture called a shooting board. When guided by a shooting board, a plane with a razor-sharp edge, set to take a light cut, can accurately slice off wispy thin shavings, as shown in the photo at left above. And the end grain will be left with the smoothest surface possible. To use one of these fixtures, first place a workpiece against the fence, and lay a handplane on its side with the sole against the edge of the base. Butt the work up to the plane sole, and then push the plane by the work in several passes.

The shooting board I use is an adaptation of an old bench hook, or sawing board. I made this combination bench hook/shooting

From *Fine Woodworking* (May 1994) 106:72-73
Photos: Alec Waters

board so it would either hold stock while sawing (see the photo at right on the facing page) or precisely plane the ends of stock. One of the fixture's unusual features is its removable 45° fence, which makes it both a miter and a right-angle shooting board. The fence is reversible as well, so I can pare miters from the left or the right side, a great advantage when I need to work each half of a joint in molded work.

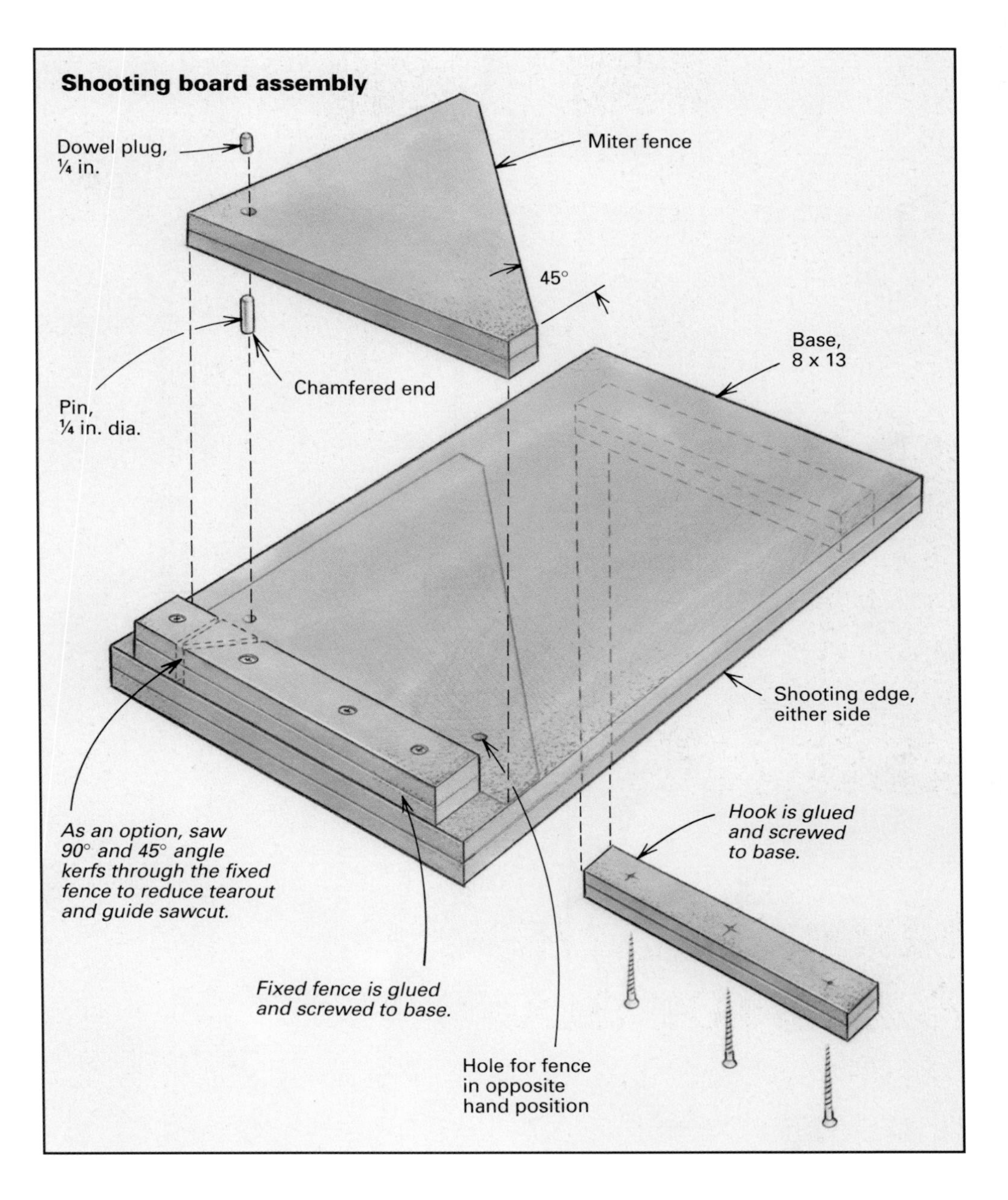

Making the fixture

My shooting board consists of a rectangular base and fence, a triangular miter fence and a hook strip, which serves as a bench stop and a clamping cleat. I made all of the parts out of medium-density fiberboard (MDF). To get the 1-in. thickness I wanted, I first laminated two pieces of ½-in. MDF, about 9 in. by 25 in. Next I cut out pieces in the sizes shown in the drawing at right, making sure all the corners were exactly square and the 45° angles were dead accurate, not just close.

When assembling the shooting board, I was concerned about how much pounding the fixed fence would take. That's why I both glued and screwed it to the base. I attached the hook the same way. First I drilled and countersunk the screw holes. Next I aligned each piece with a square and glued and clamped it to the base. Then I fastened each in place with bugle-head drywall screws.

The removable miter fence registers against the fixed fence and is held down by a snug-fitting pin. I used a ¼-in. bolt with the head cut off for the pin. As an alternate, a hardwood dowel would work, but I suspect over time the pin would become loose. Because the location of the pin and the size of its holes are critical, I bored the holes with my drill press. First I drilled a ¼-in. pin hole through the miter fence in the location shown in the drawing. Next I clamped the fence to the base in its right-hand position, so I could drill through the pin hole into the base. I flipped the miter fence and did the same thing to make the hole for the left-hand position. I chamfered the end of the pin and then tried its fit in the base holes.

Using cyanoacrylate glue, I secured the pin in the fence hole, letting the chamfered end hang out about ½ in. on the underside of the fence. For aesthetic reasons, I plugged the top ¼ in. of the fence hole with a dowel. With the shooting board together, I clamped it in my bench vise. Then I laid my plane on its side and took a shaving off the shooting edge, both sides. Because a standard plane iron does not go all the way across the sole, the iron leaves a rabbet along the base. This is necessary for proper registration of the plane. After dusting the fixture off, I finished the whole thing with oil. After it was dry, I waxed the shooting board to keep it slick and clean.

Shooting square cuts and miters

To use the shooting board, clamp its hook in an end vise to keep the fixture stable. Make sure your bench is dead flat, or lay down a flat auxiliary table before clamping the fixture. While steadying the workpiece, hold the plane with a firm grip, and keep it tight against the edge of the shooting board as you take multiple passes. Use the largest bench plane you have. A Stanley #7 or #8 jointer plane works best, but a #5 jack plane will also do, as long as it has a sharp iron, squarely set, and its sole is true and square to the plane's body. Even better, you can use a miter plane, which resembles an oversized block plane and is specifically meant for shooting (see the photo at left on the facing page).

When shooting the end grain of a right angle cut, it's a good idea to knife an edge line around the board, which will prevent tearout, and then plane to the line. When shooting 45° angles, tearout is rarely a problem. In this mitering mode, the shooting board can trim tiny amounts (see the photo at left on the facing page). This is crucial when fitting a lipping around a veneered panel, for example, where the length of the lipping from inside miter to inside miter has to be exactly the length of the panel. Because the fence pin serves as a pivot point, you can adjust the angle of cut slightly to bisect a corner that's not quite square. Just insert a paper shim where needed between the fences. I have a stack of old business cards that work great for this. □

Ed Speas is a woodworker in Ballground, Ga.

Drawing: Bob La Pointe

Scratch Awl from Scrap

Simple steps produce a beautiful, high-quality tool

by Tom Herold

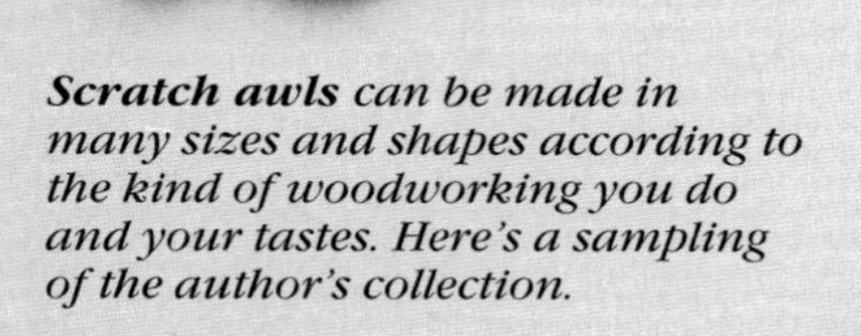

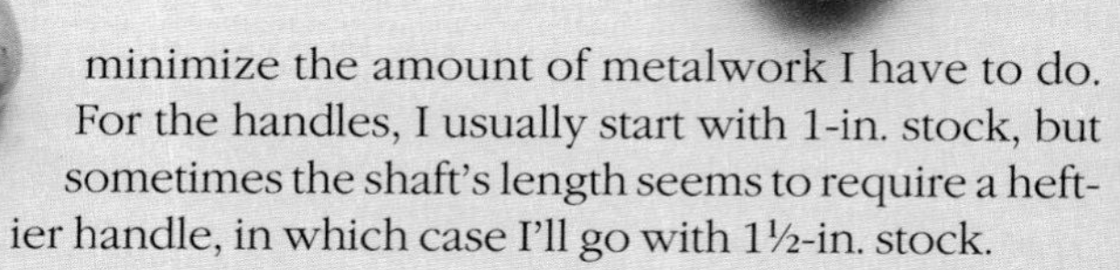

***Scratch awls** can be made in many sizes and shapes according to the kind of woodworking you do and your tastes. Here's a sampling of the author's collection.*

A scratch awl (or scribe) is an indispensable marking tool, which is capable of striking a finer and more useful line than you can get with a pencil. Beautiful versions of the tool are available commercially but often cost upward of $35. For about a tenth of that, you can make one of your own. The tools you'll need are all fairly common: a woodworking lathe, a 3-jaw or 4-jaw chuck, a drill press, a grinder and a standard propane torch. Once you've made your first scratch awl, and you see how simple the process is, you'll make many more. Besides being fun to make, a scratch awl you've crafted yourself, which can't be matched by any tool you can buy, is satisfying to use.

Making a scratch awl is a great first project in metalworking, but you need to be aware of the hazards. When cutting metal on a wood lathe, remember to use eye protection, keep your hands and clothing out of the way and concentrate on the task.

Selecting materials

Most of the awls I've made have been between 5 in. and 8 in. overall. I like to size my awls to take advantage of standard material sizes, most notably 3⁄16-in.-dia. steel rod and ½-in.-dia. brass rod, to minimize the amount of metalwork I have to do.

For the handles, I usually start with 1-in. stock, but sometimes the shaft's length seems to require a heftier handle, in which case I'll go with 1½-in. stock.

For these awls, I used O-1 steel (a high-carbon, oil-hardening tool steel), which can be purchased through many industrial-supply companies. It costs only a few dollars a linear foot. I bought my brass at a scrap-metal yard for $2 a pound. I've also seen brass rod at home-improvement stores, but it's much more expensive. Making these awls also lets me use some of those beautiful scraps I can never throw away, and even if I buy handle stock, I can buy "shorts" from lumber dealers for very little and have the experience of working with an otherwise unaffordable exotic.

Working with metal

Metalworking isn't that much different from working with wood; the material's just harder. I begin by hacksawing a piece of steel rod about 5 in. or 6 in. long and chucking it in my lathe's headstock, making sure the steel protrudes about ½ in. Using the lathe's slowest speed, I file the end of the steel smooth and flat to

Photo: Susan Kahn

ready it for end drilling. Next I chuck a ¼-in. center drill (available from industrial-tool suppliers) into the tailstock of my lathe, squirt a bit of oil on the end of the steel rod and bore a hole in the end, just deep enough to seat the tailstock center, which will support the steel during turning.

Next, to prepare the brass collar, I cut a piece of ½-in.-dia. brass rod ½ in. long and chuck it in the headstock. The brass provides a nice transition from steel to wood. I clean up the end of the brass with a file and then use a skew to get the end flat where it will meet the wood. As with the steel, I use the lathe's slowest speed. After squirting a couple of drops of oil where I'm drilling to lubricate and cool the cut, I center drill the brass to the same diameter as the steel shaft.

I remove the brass from the lathe, clean it and the steel *thoroughly* with lacquer thinner and slide the steel through the brass. It's essential to remove all oil and dirt from both steel and brass; if you don't, you won't get a good solder joint. I leave enough steel on the handle side of the brass to form a tang, which will seat well in the handle. I leave enough steel on the other (center drilled) end for the shaft plus a little extra, which I'll cut off after tapering the shaft. I use a propane torch and regular pipe solder, making sure I get a good flow of solder on both sides of the brass. I don't worry about any excess solder now because I'll clean it up during the next operation.

I rechuck the brass in the headstock and support the center-drilled end of the shaft with the tailstock. Then I clean the solder joint with a metal file and shape the brass with a skew. Next I taper the steel shaft with a fine mill file, leaving just enough metal at the point for support—usually about 3/32 in. thick. This seems to take forever, but it's really only about 10 minutes. I sand next, from about 220-grit down to 2,000-grit (very fine abrasive papers are available at most auto paint shops), which gives the shaft a nice finish.

I reverse the awl in the chuck (chucking the brass collar), so the tang is exposed. I clean the tang side with a file and skew just as I did the other side, making sure the brass is flat and perpendicular to the steel rod, to ensure that the handle seats at the collar. I rough up the tang with a file to promote good adhesion and file a portion of it flat to ensure the handle won't rotate on the shaft.

Making the handle

When picking wood for handle material, I look for interesting figure, dramatic color or just plain beautiful wood. To prepare the handle blank for turning, I get one end flat and smooth, and then I drill a hole in the middle of that end deep enough for the tang and about 1/32 in. larger in diameter. The extra space prevents the joint from being glue starved. After cleaning both steel and brass with lacquer thinner, I epoxy the handle to the shaft. If I can spare my lathe for 24 hours, I'll do it with the awl still in the chuck. Once the epoxy has cured, I turn the awl's handle to its final form.

I take it slowly when turning the handle. Although it's not easy to break the glue joint, it is possible. Sometimes I use the tailstock to keep the handle turning true until it's pretty much roughed out. Once the handle is close to the desired shape, I remove the brass collar from the chuck, back the awl out a bit and chuck the shaft. This gives me room to clean up the brass and get a good transition from the handle to the brass.

Once I'm satisfied with the handle's shape, I sand it smooth and finish it. I've used shellac, linseed oil and Formby's tung oil finish. It all depends on what look you like. My dad insists that the best finish for a wooden tool handle is just plain wax. I can't argue with that. With the handle finished, I take the awl out of the chuck, put the center-drilled end of the shaft in a vise and, using a hacksaw, cut off the tip that extends beyond the taper.

Tempering and sharpening the shaft

The final steps in making the awl are tempering and sharpening the end of the shaft. Tempering is a two-stage process: hardening and drawing. To harden the steel, use your propane torch to heat the end of the awl to a cherry-red (or glowing red) color, just after it's gone through dull red. I hold the awl by its handle with one finger just touching the steel near the handle, and place the end of the shaft (back just a bit from the point) into the tip of the flame. As soon as the end of the shaft becomes cherry red (and before the shaft gets too hot to touch), I quickly place the shaft into a nearby can of motor oil.

Quenching the steel in oil like this will bring the temperature of the steel down rapidly, making it extremely hard and brittle—almost like glass. Because it's far too fragile for use at this point, I have to remove, or "draw," some of the brittleness from the steel by heating it up again. But this time I heat it only until it reaches a light straw color—about 430° Fahrenheit. Before I heat it up again, though, I clean up the shaft with very fine abrasive paper—1,000-grit—so I can see the color of the shaft when I do reheat it. Then I position the tip of the awl well above the flame and move the shaft, in and out of the heat. Once the shaft starts to change color, the process goes very quickly. So carefully watch for the steel to start to take on the light straw color, and be prepared to plunge the blade into the motor oil immediately.

If all this talk of *cherry red, dull red* and *light straw* sounds a bit daunting, don't worry. Temperature-indicator cards are available wherever welding supplies are sold. They show spectrums of color in correlation with temperatures and allow you to hazard a fairly accurate guess as to the temperature your steel has attained.

Finally, I sharpen the awl's shaft on a grinder. It's easy and virtually foolproof, just like sharpening a pencil on a belt sander. But then again, none of us have ever done that. □

Tom Herold is an aerospace engineer who works wood for pleasure. He lives in Colorado Springs, Colo.

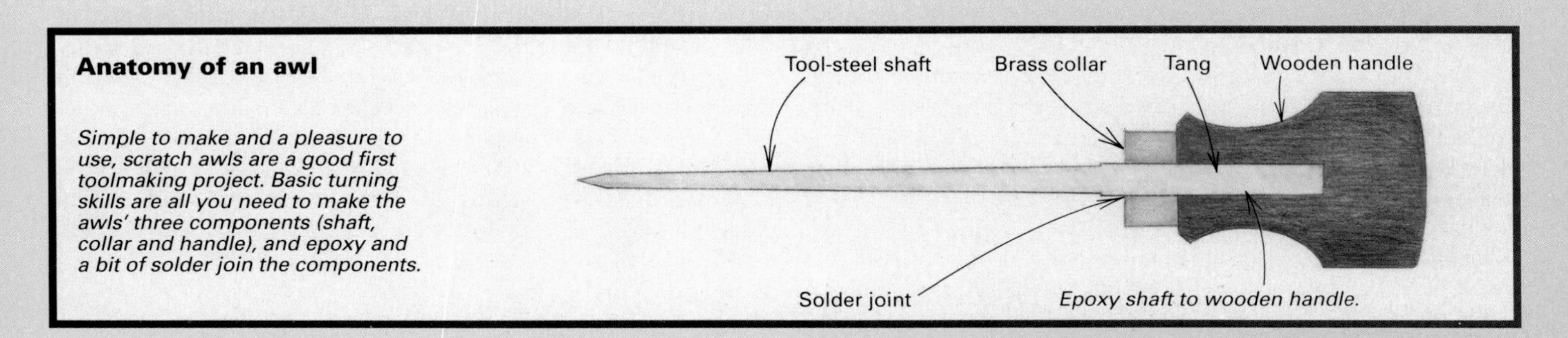

Anatomy of an awl

Simple to make and a pleasure to use, scratch awls are a good first toolmaking project. Basic turning skills are all you need to make the awls' three components (shaft, collar and handle), and epoxy and a bit of solder join the components.

From *Fine Woodworking* (May 1993) 100:56-57

Making a new tool handle *doesn't take a lot of time, and the benefits are well worth it. A new handle (see inset) will make the tool more comfortable and easier to control, contributing to better woodworking.*

Get a Grip on Your Tools

Making and modifying handles for comfort and control

by Christian H. Becksvoort

I had my dovetail saw for some time before I gave it its first real workout. I was making a large case piece with almost 400 hand-cut dovetails. Halfway through the first day of sawing, I began to notice blisters forming. I realized my high-priced, brass-backed, British dovetail saw needed some serious customizing if I was ever to finish the piece. Out came the knife, rasp, file, sandpaper and gouge. I then chopped and carved the handle until it fit my hand. A little more scraping and sanding, followed by a coat of oil completed the project.

In about an hour, I had converted the stock saw handle from a painful tool into one that I look forward to using because it fits like a glove. But if your stock handle is beyond hope or just plain ugly, another option is to start from scratch and create a unique custom handle for your saw, as discussed in the box on p. 94.

Tool handles are the link between you and the tool. This is where control takes place. If the hand is not perfectly comfortable and at ease, then you lose some of that control. Consequently, a well-fitted handle can make you a better woodworker.

That saw was my first venture into customizing handles, which has since spread throughout my toolbox (see the photo on p. 93). Since that time, I have modified or replaced handles on tools such as chisels, planes, drills and braces, clamps and screwdrivers.

The decision of whether to modify or replace is an individual one. I have an aversion to plastic handles. They get replaced immediately. I also dislike garish bright paint and tinted lacquer. For instance, I scraped and sanded off the finish on my wooden handled screwdrivers, and in the process, I also filed and sanded their rough ends. Carving tools received the same treatment. Anyone who has ever held a gouge for more than 10 minutes can appreciate the comfort of a smooth, rounded handle end against the palm.

I couldn't find a wooden handled, square-drive screwdriver, so I salvaged the wooden handle from a worn-out Phillips-head screwdriver. The wooden handle pulled off the shaft relatively easily, but I had to cut the plastic handle off the square-drive screwdriver using a worn-out blade on the bandsaw. I then drove the square-drive's shaft into the salvaged wooden handle.

Photos except where noted: Charley Robinson

Rehandling planes

Years ago, planes had rosewood handles and knobs, then walnut, then stained beech or birch, and now, many handles are plastic. The rosewood handles I leave alone, and I usually don't change walnut handles unless they're damaged. But I do replace the others. I usually turn a new front knob first. The existing knob is a rough guide, but now's the time to adapt it to your grip. The only fixed diameter is where the knob mates to the plane body.

To drill the centered, counterbored hole into the top of the round knob, I first clamp a piece of scrapwood to the drill-press table. Then I drill a flat-bottomed hole, the same diameter as the knob base, about ¼ in. deep into the scrapwood. I change bits to fit the bolt head, place the knob in the hole and drill the countersink for the bolt head. The depth is critical to get the bolt perfectly flush with the surrounding knob. Finally, I change bits again and drill for the bolt shaft. Then all the knob needs is a little oil, and it's ready to screw onto the plane.

The back handle or tote is just as easy. I bandsaw the rough shape, drill the countersunk hole for the long bolt and then shape the handle to suit with a rasp, file and sandpaper. I carved the last handle I made, as shown in the photo on the facing page, and left the facets (see the inset photo on the facing page). I liked the look of the carved handle, and I thought the hewn texture would improve the grip. Finally, I drill the hole for the short front-mounting bolt. After a coat of oil, I attach the handle to the plane. As with other handles, I add a few more coats of oil over the next few days.

Turning chisel handles

My first effort involved a set of Greenlee socket chisels, which I reshaped by chucking the tapered end of the handle into a three-jaw chuck on my lathe. I turned the handles into smooth cones with a gentle curve and sanded and finished them in a few minutes.

New socket handles are almost as easy to make. I cut my rough stock to 1½-in.-sq. blanks, 6 in. long and place it into my toaster oven (set to 200°F) for one or two days. This dries the wood to 0% moisture content. Then I turn the blank to shape and cut a taper on one end to fit the chisel's socket. I turn the other end for a stainless-steel hoop to prevent the handle from splitting when hit with a mallet. I prefer stainless steel to brass or copper for hoops because it doesn't tarnish, and the silver color goes well with any wood. I buy ¾-in.-ID stainless pipe at a local machine shop, have them cut it to ⅜-in. lengths and radius the edges of the resulting hoops. I then polish the hoops before fitting them to the handle.

It's important to turn the handle to fit the hoop immediately after drying and to have the hoops on hand to test the fit. If you get a good, snug fit at this stage, when the wood returns to equilibrium moisture content, it will swell and securely anchor the hoop.

Tang chisels are even easier. I select the stock and drill the hole for the tang. The hole should be undersized to get a good fit. The only critical dimension on the turning is at the bottom to accommodate the ferrule, either salvaged or new, that keeps the handle from splitting when the tang is forced into the handle.

Split-handle installations

My most challenging handle replacement was for a set of Stanley #40 chisels. The handles on these chisels consist of a metal strike cap on the top of a 1-in.-dia. metal shaft that was forced into the top of a hollow polycarbonate handle. The chisel shaft was forced into the other end of the handle until it met the strike-cap shaft. It does not require an advanced degree in metallurgy to figure out that two pieces of steel shaft forced against each other by constant pounding will eventually mushroom. Indeed, one handle in my set would split every three to six months. I would send the chisel back to Stanley and they would dutifully send me a new one.

They stopped making the #40s in the late '70s. At that point, I bandsawed the plastic handles off and pondered a solution. I had a machinist friend and neighbor turn new strike caps with long, oversized shafts, which he drilled out to accept the chisel shafts. I took a wooden blank, cut straight at both ends to the correct length. Then I drilled it from the top to accept the strike cap sleeve and from the bottom to accept the remaining chisel shaft. Next I turned the handle to approximate the old plastic handle. After oiling the hollow handle, I took a wide chisel and split it lengthwise. Finally, I epoxied and clamped the two halves around the shafts of the chisel and strike cap. With a little more sanding and oiling, the split line is virtually invisible. I used the same split handle technique to replace the plastic handle on the swing arm of a brace after bandsawing off the old handle.

Rehandling machines

Wooden handles on stationary power tools are a pleasure to use. For odd sized bars or control levers, I usually drill the mounting hole into my wooden handle blank the next larger size, turn the knob and epoxy it into place. For screw-on knobs, I use threaded brass inserts of the correct size, fitted into the knob. □

Christian H. Becksvoort builds custom furniture in New Gloucester, Maine, and is a contributing editor to FWW.

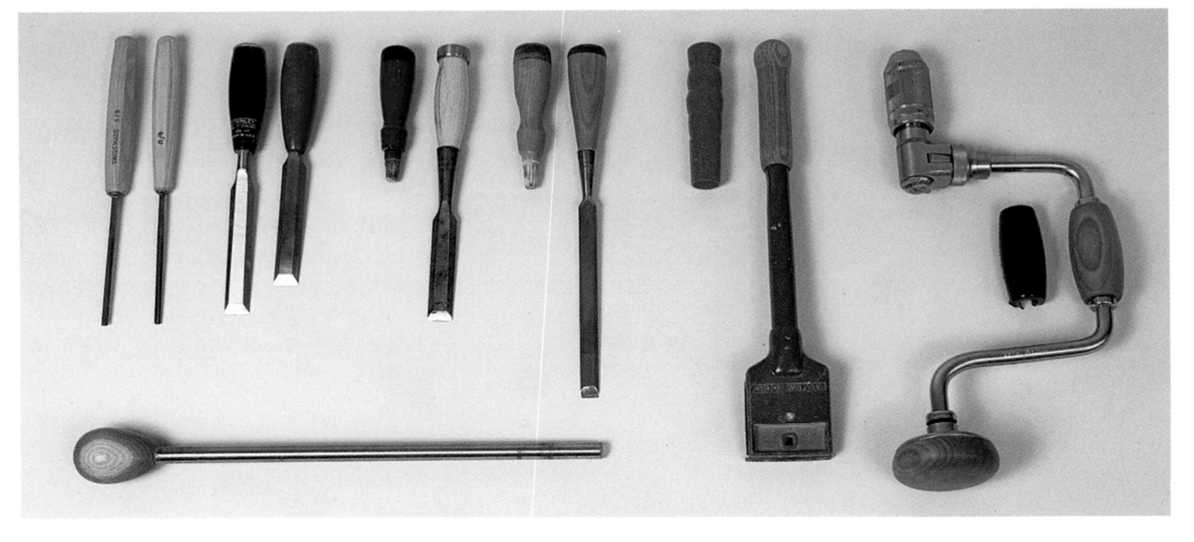

Rehandling a shopful of tools—*Shown here are before (on the left) and after examples of tools that have modified or new handles. From the left, Pfeil carving gouges (edges eased and sanded), Stanley #40 chisels (handle replaced, new steel cap), Stanley #750 socket chisels (handle replaced, stainless steel hoop), Greenlee chisel (handle reshaped), Allway Tools scraper (handle replaced), Stanley #923 brace (handles replaced), Shopmade lathe knock-out bar with black locust knob.*

From *Fine Woodworking* (March 1994) 105:41-43

Regrip your saws

by Mario Rodriguez

We've all admired the beautiful handles on antique woodworking saws. Their flourishes, rooster tails and swirling cusps suggest movement and speed. Usually made of beech, apple or pear, these handles are minor works of art. Today, most saws are fitted with blister-raising slabs of wood that cramp the hand and fatigue the arm.

A custom handle will improve the performance as well as the look of your saw. A comfortable and properly shaped grip makes it easier to guide the saw for more accurate cuts and excellent results.

Choosing your material: Handle making requires only a small piece of wood, so you might as well use some wildly spectacular stock. I've salvaged most of my handle stock from the scrap bin or firewood pile. Start with a piece of ⅞-in.-thick, unusual or figured wood that's about 6 in. by 7 in., orienting the grain lengthwise.

Choosing a handle pattern: Small backsaws have pistol grip handles. Large backsaws and full-length handsaws usually have a closed or hollow handle. I have six handle patterns I use (see the photo at right), and I'm always looking for new ones. I frequently make a tracing of interesting designs. Sometimes I copy a design from an old tool catalog and have it enlarged. Experiment and anticipate changing the handle more than once until you find a grip that works for you. A good design evolves after some use and time for evaluation.

A pistol grip pattern is easy to make and will work for most small backsaws measuring 8 in. to 12 in. long. The shape is easy to cut out. There are no interior cuts, and it can be done on the bandsaw.

A closed or hollow handle requires an interior cutout. First drill pilot holes, and then complete the cut on a scroll or jig saw. Or you can cut the blank into two pieces, make your interior cutout, and reglue and complete your exterior silhouette.

Shaping the handle: You can quickly round corners using a table-mounted router with a ball bearing-guided bit. Taking several shallow cuts, round the interior cutout and the back of the grip. Don't eliminate the flat surfaces. A handle that is too round tends to be slippery. For the rooster tail, I prefer a thumbnail profile cut by hand with chisels and files. Try for crisp, sharp edges.

Plotting the cut for the sawblade: Lay the sawblade onto your handle. Mark the top and bottom corners of the sawblade onto the handle, and draw a line connecting the two. This should be the baseline of the cut into the handle for the blade.

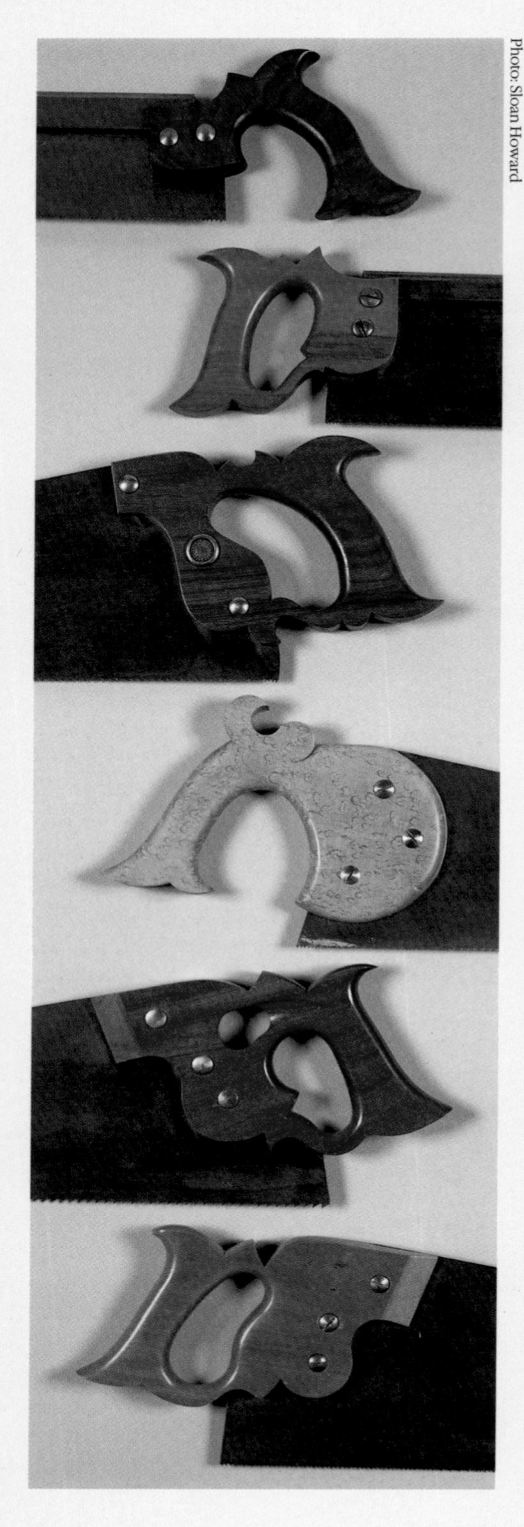

Photo: Sloan Howard

A special jig holds the handle blank (above) when bandsawing the sawblade kerf. Although a bandsaw kerf is slightly oversized, it doesn't affect the final fit of blade to handle.

***Old patterns for new handles**—The author finds patterns for his new hardles by tracing directly from an old saw that he likes or in old catalogs (left).*

I usually pitch the handle forward at 50° to 60° in relation to the cutting edge, as opposed to the stock angle of about 70°. This helps me to direct the force of my stroke directly behind the blade for easier, more controlled cutting.

Cutting the blade slot: For speed and accuracy, I usually cut the slot on a bandsaw fitted with a ¼-in.-wide, 6-teeth-per-inch blade. Although a bandsaw leaves a kerf wider than the thickness of the handsaw blade, the difference is negligible and ultimately will not affect the fit of the blade. For safety, I use a jig (see the photo above right) to securely hold the handle with the blade slot baseline parallel to the bandsaw blade.

I sometimes cut the slot by hand for a tighter fit. I use the saw to be rehandled with the old handle reattached temporarily. However, this method requires extreme care because if your cut is off line, it will force a kink into your blade when you fit it to the handle.

Spine mortise: If you're working with a backsaw, you'll have to cut a narrow open-ended mortise into the top of your handle to accommodate the spine. Mark on each side of the blade slot for the width of the spine, and carry the mark down the front of the handle. Cut the sides of the mortise with a small dovetail saw, and then carefully, pare out the waste with a ¼-in. chisel. Periodically, set the blade into the handle to check for fit. Cut this mortise on the small side and enlarge as necessary to avoid any gaps in the finished job.

Drilling the holes: Make a cardboard template of the heel end of the blade, including the blade holes. Line up the template to the blade baseline drawn on the handle. Drill the screw and nut holes slightly oversized for an easy fit. For a cleaner look, you should countersink the saw nuts almost flush with the handle.

Finishing the handle: After sanding with 320-grit paper, I spray a light coat of lacquer sander/sealer, followed by a coat of gloss lacquer. Sometimes, I brush on two coats of shellac instead. When dry, rub the finish with "0000" steel wool, then wax. □

Mario Rodriguez is a contributing editor to FWW *and a cabinetmaker, teacher and woodworking consultant.*

Toolbox Tours de Force

Tool-storage solutions show style, ingenuity

by Vincent Laurence

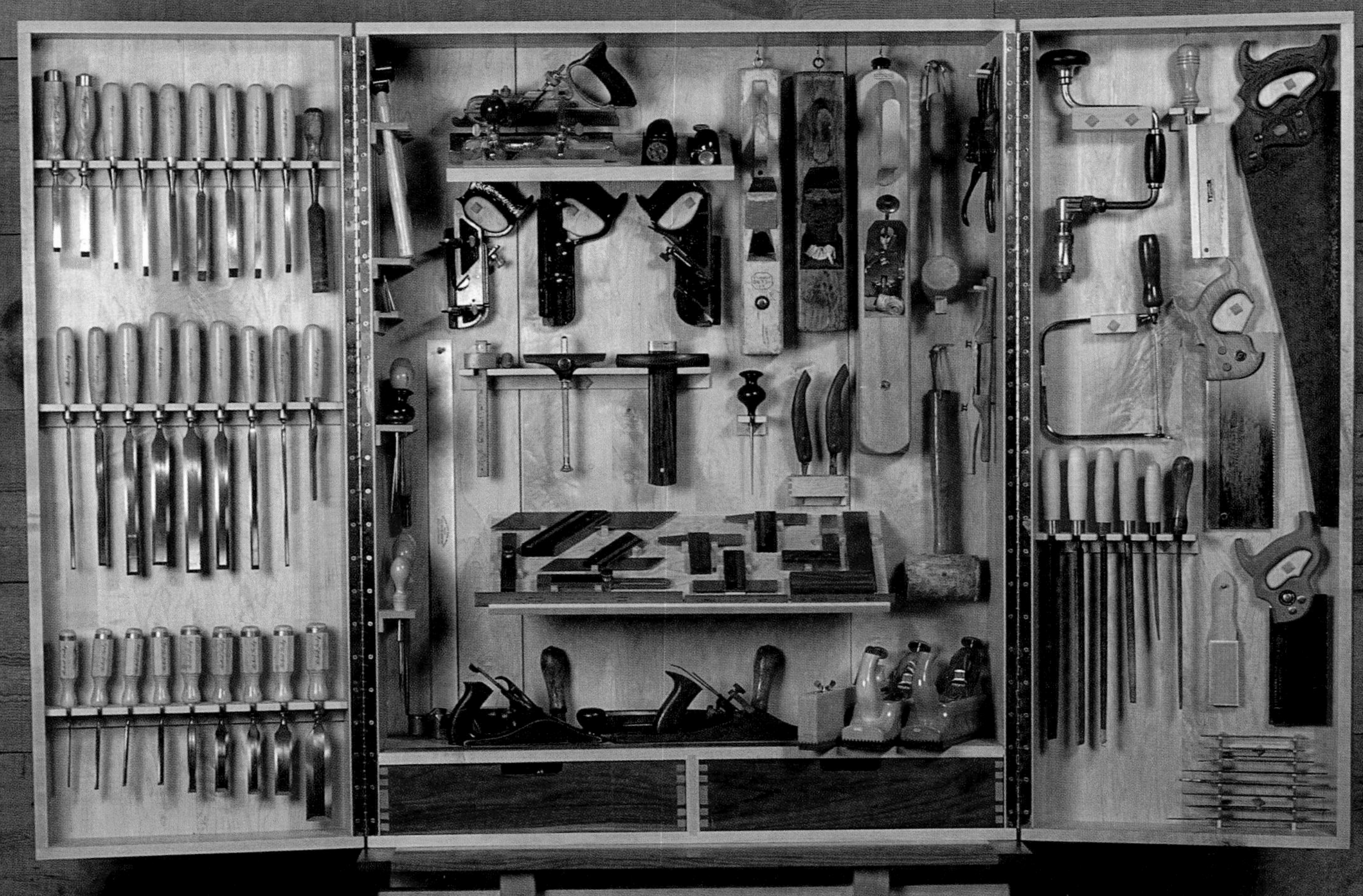

***Convenient access to their tools** was a prime design consideration for Herbert Buchalter and his son Sanford of Freeland, Mich. Father and son share a shop and tools. The younger Buchalter, a professional furnituremaker, built the cabinet for his dad, though it's the son who gets the most use of it.*

To many woodworkers—even some of the best—a simple, unprepossessing plywood box, biscuit-joined together, is perfectly satisfactory for tool storage and organization. The materials and construction are in keeping with the function of the box—after all they're just tools, right?

Other woodworkers see things differently. To them, the toolbox, though still primarily functional, must also be beautiful, something that affirms why they do what they do, day in and day out. Many of these woodworkers also see their toolboxes as three-dimensional portfolios on view for prospective clients, testifying to the skill of the maker.

When we asked readers to send us photos of their toolboxes in "Editor's Notebook" (*FWW* #100), we weren't disappointed. From the more than 30 boxes, chests and wall cabinets readers sent photos of, we selected nine of the finest (eight are shown here; the ninth is on the cover of *FWW* #105). □

Vincent Laurence is an associate editor of Fine Woodworking.

Photo: Jonathan Binzen

This traditional 19th-century-style chest *(above) houses over 400 tools, weighs more than 300 lbs. loaded with tools, and took its owner, Tony Konovaloff of Bellingham, Wash., nearly 200 hours to make. Konovaloff, a professional furnituremaker, uses hand tools exclusively, so his entire shop is in this chest. The Latin phrase carved into the lower rail of the lid means "Art is long, life is short."*

"A place for everything and everything in its place," *one of the Shaker creeds, might just as well describe Konovaloff's attitude toward tool storage. Every tool has a specific location that it fits precisely, thus keeping the box neat and the tools accessible (inset).*

With the inside of its lid flipped down, *Konovaloff's box reveals its collection of saws and extra blades (right).*

Photos except where noted: Vincent Laurence

Photos left and right: Jonathan Binzen

Steve Johnson's mobile tool chest *was built to withstand the abuse of bouncing over cracked concrete as it is rolled around a factory (he's a professional tool-and-die maker). Built of solid walnut and designed much like the automotive tool chests he was used to, the drawers on his chest glide on full-extension, 100-lb. slides and are attached to an internal carcase of ¾-in. particleboard. Overall weight for the chest, without any tools in it, is about 215 lbs. (including a 60-lb. granite surface plate mortised into the top). The machinist's-style box on top of the chest doesn't travel with the larger chest, but is a good companion in the shop.*

Case for an "artist"—*Tired of being told, "Hey buddy, use the freight elevator" every time he ventured into a high-rise apartment building for an installation, cabinetmaker Eric Sheffield traded in his old toolbox for the Gibson guitar case you see here. It holds an incredible number of tools—over 40 lbs. worth—and now, according to Sheffield, he's "accorded the respect a true artist deserves."*

From *Fine Woodworking* (March 1994) 105:53-57

***Being able to see his tools** and get to them easily was a priority for furnituremaker Greg Radley of Ventura, Calif., when he started designing this tool chest. Radley's solution was this chest-on-trestle with a utility cabinet tucked into the trestle below. The trestle and all frame components are solid ash, the panels are solid mahogany veneered with curly European ash and the interior partitions are all mahogany.*

Four-part portability—*"If I can't move it, I can't have it" is the imperative that guided the design of Harold Purcell's toolbox (right). His solution, a maple base and three stackable, cherry and mahogany boxes, provides Purcell with convenient access to his tools as well as a fair measure of visual and tactile satisfaction.*

Tim Kimack's veneered, inlaid and entirely handmade tool chest *(below) makes a fine home for his collection of antique and owner-built tools. Kimack, a finish carpenter and furnituremaker in Simi Valley, Calif., put over 400 hours into the chest, calling it "definitely a labor of love."*

Two for the road—*Like many an idealistic young woodworker, David Sellery of Santa Cruz, Calif., had visions of Krenovian masterpieces dancing in his head when he first started working wood. Years later, he found he had built a few more kitchen cabinets than he'd preferred just to pay the bills. But Sellery made the time to build this pair of carpenter-style toolboxes to remind him of why he first started working with wood. Though almost jewel-like in their detail, they're sturdy, functional and see daily use on the job site with no apparent ill effect.*

Building a Thickness Sander

A large drum turns an abrasive machine into a smooth operator

by William "Grit" Laskin and David Wren

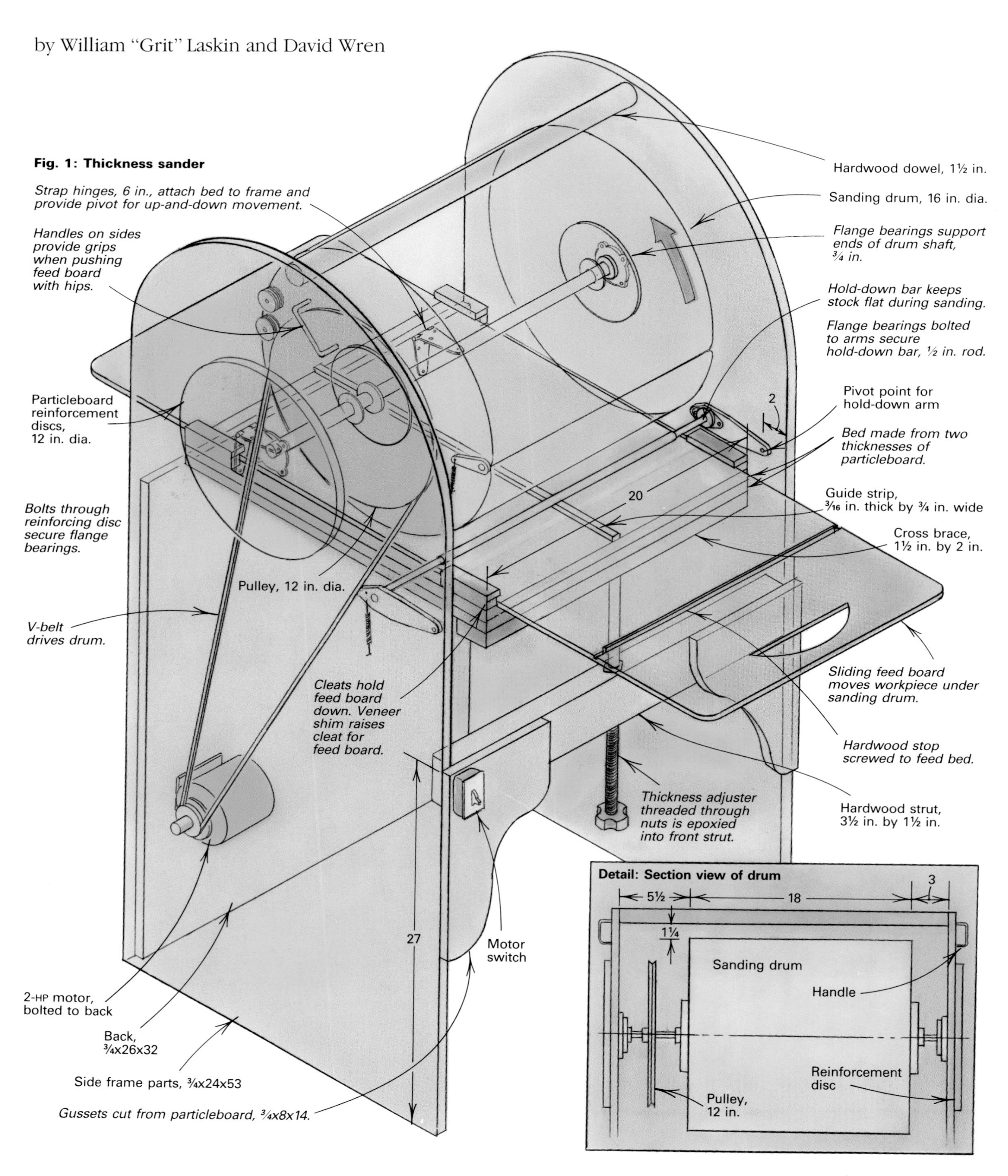

Drawings: Aaron Azevedo

Like most other woodworkers, guitarmakers routinely need to plane and smooth wide, flat pieces of wood for instrument parts. Unfortunately, the job is usually too delicate for most stationary thickness planers. As guitarmakers, we often have to plane the sides, back and top of a guitar to ⅒ in. or less, and most planers don't surface stock much thinner than ⅛ in. Further, many of the wood varieties typically used by guitarmakers, such as rosewood, curly maple or koa, are susceptible to splintering and tearout from the cutting action of a thickness planer. The alternatives are scraping, handplaning and hand-sanding the wood or running the stock through a thickness sander, if you have one at your disposal.

Thickness sanders come in several different styles. The Cadillacs of the breed are the wide-belt sanders: expensive machines that sport power feeds, accurate thickness adjustments and oscillating sanding drums. The lower-cost alternatives are drum sanders that have a small-diameter drum, wind-on sandpaper and manual feed. The problem with these is that resinous woods, like rosewood, easily clog the small surface area of the sandpaper. We considered building one from mail-order plans, but those we saw were so poorly conceived, we decided that we could design and build a better machine ourselves. The machine we came up with, shown at right, fulfilled our basic tenets: it was quick and inexpensive to build and it operates to close tolerances. Building the machine requires mostly woodworking tools and skills and very little metalworking. It will be a welcome addition to any woodworker's shop.

Anatomy—Our sander incorporates elements of both wide-belt and small-diameter drum-style machines. The large-diameter sanding drum provides a lot of surface area, similar to a wide-belt sander, which extends sandpaper life. But because the single sheet of paper is wrapped around the drum, there's no need for an elaborate belt-tracking mechanism, as found on the wide-belt machines.

The large-diameter drum is made from separate discs that are cut and routed to shape and then glued together. A special grip mechanism, which will be described later, secures and tightens the sandpaper around the drum. To change sanding thickness, the sander's bed moves relative to the drum, not the other way around. The bed is hinged at the back end and rests on a threaded rod at the front, allowing adjustment. The workpiece is supported and guided by a feed board that slides along the bed. The feed board is moved manually, eliminating the need for a separate power-feed drive mechanism—keeping complex construction to a minimum.

To keep costs down, most sander parts are ¾-in.-thick particleboard. Other supplies needed for building the machine include flange bearings, a few pulleys and a 2-HP motor, all available from W.W. Grainger Inc., 5959 W. Howard St., Niles, Ill. 60648; (312) 647-8900. You'll also need strap hinges, springs and a few other supplies, which should be available from your local hardware or building-supply store.

Building the frame—The body or frame of our thickness sander is made up of nine pieces: two side panels; one back panel; a front strut and two reinforcing gussets; a dowel strut at the top edge; and two circular side reinforcements. All the pieces were cut to the dimensions given in figure 1; straight edges were cut on the tablesaw and curves on the bandsaw. The pieces were then assembled with glue and screws run into predrilled holes. The most sensible order for assembly is to first attach the reinforcing circles to the side panels, and then attach the back panel to the sides. Next take the front cross strut and mark the location for the threaded rod that will raise and lower the table. Drill a hole larger than the rod through the strut, and use a ½-in. chisel to mortise a space to inset nuts for the rod on both the top and bottom of the strut. Glue the nuts in place with five-

A large-drum thickness sander can provide even a small shop with the capacity to abrasive-plane and smooth wide stock. This sander, designed and built by Grit Laskin and David Wren, is economical to make: the drum and most parts were cut from ¾-in.-thick particleboard.

minute epoxy while threading the rod through them, to make sure the threads align. When the epoxy has set, remove the rod, and glue and screw the strut to the frame, as well as the gussets to the sides and the strut. Also, glue and screw in the large dowel that reinforces the sides above the drum. Finally, attach a handle to the threaded rod. A simple block of wood is fine; you just want something that will enable you to turn the rod in small, incremental movements.

Bed and feed board—To ensure that the bed would be sturdy enough to resist the pressure of the drum without deflection, we made it from two pieces of particleboard glued and screwed together. Further reinforcement is provided by a thick hardwood cross brace that spans the infeed side of the bed and provides a bearing point for the adjustment rod. The photo above shows an additional curved piece of mahogany beneath the bed. Ignore this; it was an experiment on the first sander we built. You should locate the strut as shown in figure 1.

A feed board that directly supports and guides the workpiece during sanding rides atop the bed. The feed board is a piece of ¾-in.-thick particleboard that's a little less than 1½ in. narrower than the bed and long enough to support the workpiece, as well as engage the hold-down cleats after the workpiece is through the drum. For our sanding needs, a 53-in.-long feed board is long enough to support guitar side strips, which are normally 32 in. to 34 in. long.

To guide its travel along the bed, a ¼-in.-deep by ¾-in.-wide groove is dadoed along the center of the feed board's bottom. The groove accepts a ³⁄₁₆-in.-thick guide strip that's screwed to the bed. Two cleats at both ends of the feed board capture it and keep it flat on an extended bed. To ensure smooth feed-board movement, a

From *Fine Woodworking* (November 1990) 85:80-83

To ensure that the sandpaper wears evenly, narrow workpieces are run through the sander on an angle. The spring-tensioned hold-down bar, just in front of the drum, puts downward pressure on the stock, to help keep it flat during sanding.

thin veneer shim between the cleats and the bed provides a bit of clearance. To keep the workpiece from slipping backward during sanding, a 3⁄16-in. hardwood stop (we used ebony) is inset into a 1⁄8-in.-deep groove in the feed board (see figure 1 on p. 100). The sandpaper sheet that you'll glue to the bed later, for drum truing, also helps keep the work in place during sanding. Finally, a semicircular slot cut into the feed board provides a hand grip for hauling it back after each pass.

Install the bed to the frame with two 6-in. strap hinges, bolted through the back of the sander. While marking the hinge holes, keep the bed square to the frame and as level as possible. The cross brace at the bed's infeed end should rest on the adjustment rod. Thread the rod in place, and where it contacts the brace, screw on a small square of Plexiglas or metal to reduce wear.

Making the drum—For convenience and economy, the sanding drum is also made from particleboard. To keep the drum from being physically unwieldy, yet still big enough to yield a large sanding surface, we chose a diameter of 16 in. This translates to a more than 4-ft. circumference, which provides an ample sanding surface. Because we needed to be able to sand a large guitar back or top, which can be more than 16 in. wide, we made the drum 18 in. wide.

The cylindrical drum is made by stacking 24, 3⁄4-in. particleboard discs, which are individually rough cut, template routed, and glued and screwed together (see figure 2 on the facing page). Later, the sandpaper clamp is added, and the drum is balanced and trued to the bed. But the first task in making the drum is to bandsaw each disc (a sabersaw is also good for this job), making each one slightly larger than its final diameter. Each disc is then trimmed to shape, including the sandpaper slot, using a piloted straight bit in a router and a 1⁄4-in.-thick Masonite template temporarily screwed to the disc to guide the cut.

To reduce the weight of the drum, we hollowed out most of the discs by sabersawing away all but a 2-in.-wide border around the perimeter of the disc and around the sandpaper-grip slot. Don't worry about making perfect cuts; the drum will be balanced later. The two outer discs and one in the center are left solid and are concentrically drilled to fit the drum's 3⁄4-in.-dia. steel shaft. The drum attaches to the shaft via a pair of 6-in. pulleys screwed to the outside discs. (Pulleys are available from a hardware store or from Grainger.) Drill three mounting holes through each pulley (easily done with a twist bit), slide the pulleys on the shaft, and use the holes as a template for drilling pilot holes. Then screw the pulleys on.

Now you're ready to glue and screw the discs together one at a time. Align the sandpaper slot and the outer edges of each successive disc so that the stack is as cylindrical as possible. The two outer discs, with pulleys attached, are glued on last. As a final touch, slightly recess the leading edge of the drum around the sandpaper slot (see the detail in figure 2). Using a rasp or file, gently round the first 1⁄2 in. on the leading edge of the slot, to prevent premature wear in that area.

The sandpaper grip—To easily attach or remove the cloth-back sandpaper that wraps around the drum, we devised a simple sandpaper grip. Start with a 1-in. hardwood dowel that's at least as long as the drum. Plane one side of the dowel flat, until you're almost halfway through, and then drill a row of small pilot holes on the leading edge of the dowel flat for 3⁄4-in.-long brads, which keep the sandpaper from slipping. Nip the brad heads with wire cutters, leaving 1⁄8-in.-high, sharply pointed studs. Trim the dowel to length, leaving 1⁄2 in. protruding from each side of the drum. The ends of the dowel are attached to lever arms cut from hardwood scraps, as shown in the detail in figure 2. With the dowel in place in the slot, screw the arms on as shown and attach one end of a spring to the bottom of each arm with an eye hook that's been pried open slightly. Stretch the springs out about one-third more than their relaxed length, mark the spot and insert a small, sturdy cup hook. Bend or slip the end of the spring around the hook and the grip is complete. To change sandpaper, you simply lift the lever arm, which releases the ends of the cloth.

Mounting the drum—The thickness-sander drum rotates on a 3⁄4-in.-dia. steel shaft (a 5⁄8-in. shaft will also work) supported at both ends by four-hole flange bearings (Grainger #5X698) bolted to the sides of the sander. The holes drilled in the sides for the bolts that hold each flange should be slightly larger than the attaching bolts themselves, to give you a bit of flexibility in aligning the drum, should you need it.

To mount the drum, first draw a vertical centerline on the inside faces of the side panel. This helps you position the drum. Now, slip the drum onto its shaft and lock it by tightening the setscrews on the two pulleys screwed to the drum. The 12-in. pulley and the V-belt that will drive the drum can now be attached, but you won't tighten the setscrew until the pulleys have been aligned with the motor later. Next attach the flange bearings to the shaft ends by tightening the setscrews in each. With the bed level and the feed board in place, lay a piece of wood, approximately 1⁄8 in. thick, on top of the feed board. Lift the drum into position, resting it on the wood, center the flanges to your pencil lines and mark the position of the flange's bolt holes. Finally, remove the drum, drill the holes, and then realign the drum and bolt it in place.

Prior to any further work with the drum, the motor must be mounted to the thickness-sander frame and wired to a switch. We chose a 2-HP, 1,750-RPM, single-phase motor, which has more than enough power for our needs. The motor base is bolted directly to the back of the sander, and the switch (we used a standard motor switch, rated to handle the amperage of the motor) is screwed onto one of the particleboard gussets on the front of the sander. A small 3-in. pulley is mounted on the motor shaft and aligned with the 12-in. pulley, and then both pulleys are tightened on their shafts. The smaller pulley driving the larger one produces a slow drum speed, which makes the sander suitable for hand-feeding.

Balancing and truing the drum—The extra bulk of the drum around the sandpaper-grip mechanism throws the drum out of rotational balance. To correct that, we screwed two stacks of large fender washers on each end disc directly opposite the grip, a process similar to adding balancing weights to car tires (see figure 2).

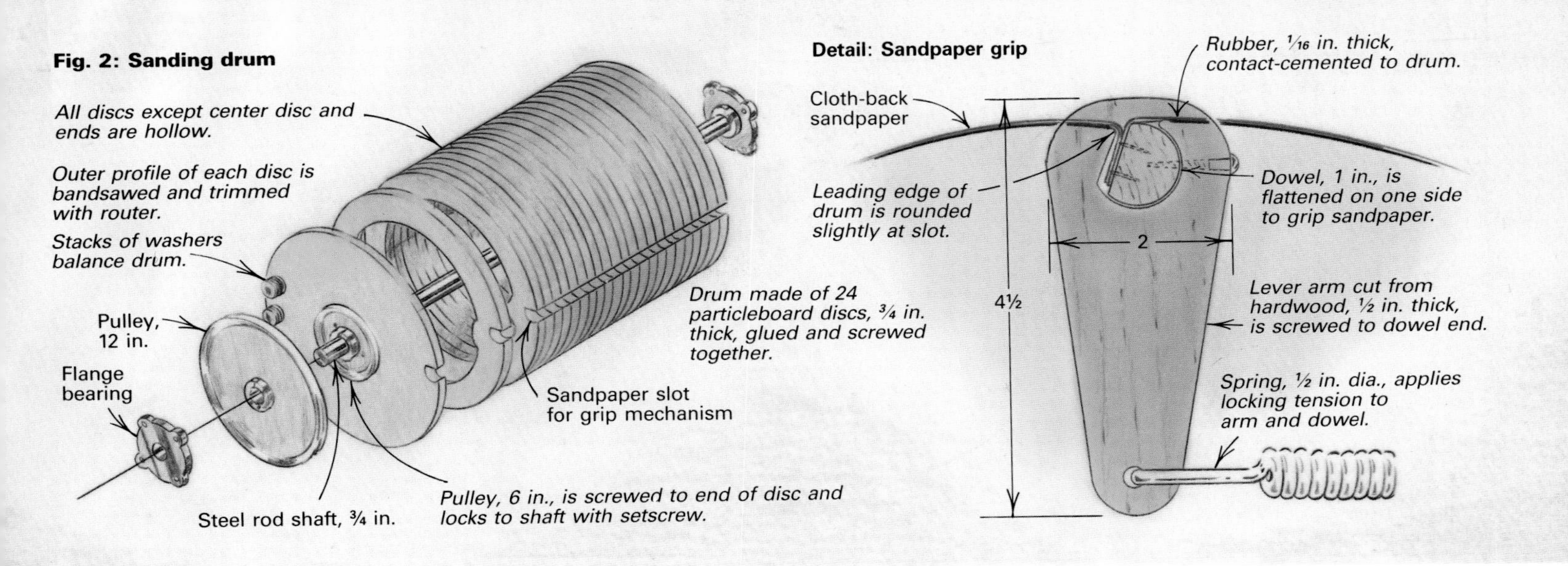

A hole drilled through the side of the frame allowed screwdriver access and made it easy for us to add weight during balancing. We just kept spinning the drum and adding washers until it came to rest at different spots after each spin—indicating that it was balanced. When you do this, make sure the V-belt is disconnected and add the same number of washers to each side of the drum; our sander took two stacks of 11 on each side, 44 washers in all.

Now comes the fun part: truing the drum. Start by contact-cementing a sheet of 40-grit cloth-back sandpaper to the feed board. With the drum under power, crank the bed up until the paper bites into the particleboard drum and sands down the high spots. Take a few subsequent passes, raising the bed and moving the feed board so that you use a fresh area on the sandpaper each time. Repeat the process until every bit of the drum has been sanded. The drum should now be perfectly even across its width and parallel to the bed. Recheck the balance, especially if you removed a great deal of material during truing.

To complete the drum, glue on a 1⁄16-in.-thick dense rubber sheet with contact cement. (For rubber sheets, check with your local building-supply store or in the yellow pages under "rubber products.") The rubber serves as a backing for the sandpaper, gives the paper a better bite and extends its life. The trick to applying the rubber is working slowly and smoothing it down as you go, to avoid air bubbles. Start with a rubber sheet that's wider than the drum and longer than its circumference; any excess can be trimmed with a sharp knife after gluing.

Hold-down bar—The hold-down mechanism is not an absolute requirement, but it's helpful, especially if you plan to sand very thin boards. The hold-down bar flattens uneven or slightly warped boards before they're sanded by the drum. The mechanism consists of a steel bar that's held loosely at either end by spring-tensioned arms that secure the workpiece tightly against the feed board as it's fed into the sanding drum. Start by covering a ½-in.-dia. steel rod with vinyl tubing, available from a conveyor-supply company. (You can substitute rubber tubing or wrap the rod with duct tape.) Leave the ends of the rod bare, as these are set into a pair of two-hole flange bearings (Grainger #4X727). These bearings are self-aligning, which means that one side can move up or down independently and not inhibit the rolling motion of the rod as uneven stock is fed through. The flange bearings are bolted to two small arms made from ¼-in.-thick Masonite that are bolted loosely to the frame, to allow up-and-down movement (see figure 1 on p. 100). Springs at each end, which pull the arms and bar down, are attached in similar fashion to the sandpaper-grip springs.

The last element you'll need to deal with before you've completed your thickness sander is dust collection. The small shroud on the original machine we built, shown in the photo on p. 101, was woefully inadequate and captured no more than 20% of the dust the sander produced. So we recommend that you equip your sander with a shroud that fully encloses the top of the machine. A piece of thin, flexible sheet metal, such as sheet aluminum (available at most hardware stores), is ideal for this job. The connection port for a shop-vacuum hose should be located on the outfeed side of the machine.

Using the sander—We've had good success using 60-grit cloth-back sanding belts. We buy this material in rolls from a local abrasives supplier; you could also buy a 36-in.-wide by 60-in.-long thickness-sander belt from The Sanding Catalog, Box 5069, Hickory, N.C. 28603; (800) 228-0000. By tearing the belt's width and length in half, you have enough for two 18-in.-wide by 52¼-in.-long sheets. This length covers the 16-in.-dia. drum (actually 16⅛ in. with rubber covering) and includes about ¾ in. extra on each end for the sandpaper grip. Secure the paper tightly in the grip, and check it occasionally during sanding to ensure it hasn't loosened. Also make sure to clean the sandpaper periodically with a regular rubber belt-cleaning stick.

To use the sander, slide the feed board out on the infeed side, and set one end of the workpiece against the stop and the other under the hold-down bar. To take advantage of the full width of the paper when sanding narrow stock, you can set the stock on the feed board at an angle (see the photo on the facing page). Now set the bed height by turning the adjustment handle. Eyeball the gap between the feed board and drum, and adjust the bed height to set the degree of sanding desired. Now switch on the motor and take a trial pass, pushing the stock through in one smooth, continuous motion. To cut down on vibration and to keep the sander from moving, we bolted the machine to a wall. Also, if your floor is slippery, you may find it easier to push the feed board with your hips while grabbing handles—regular kitchen-type drawer pulls—screwed to the frame. Beyond that, you now only have to deal with the particular idiosyncrasies of your own sander. All homemade tools have their own personalities, and once you learn their quirks, a good working relationship will quickly follow. □

Grit Laskin is a guitarmaker, author and musician. David Wren is a guitar restorer and builder. Both live and work in Toronto, Ont., Canada. A kit for the drum sander is available from W.G. Laskin, 192 Dupont St., Rear, Toronto, Ont., Canada M5R 2E6.

A Shop-Built Shaper

Tilting table adds a new angle to panel-raising

by Frank Perron, Jr.

When I was faced with the task of making a bunch of raised panels for several large jobs, I wanted to avoid the drudgery of sanding out sawblade marks on panels raised on the tablesaw. This tempted me to buy a shaper big enough to handle panel-raising. I remembered a mammoth cast-iron shaper that an old furnituremaker in Vermont had shown me. The huge machine barely vibrated as it ran, even though the raised-panel cutter mounted on it was half a foot in diameter. Further, the machine was used primarily for panel-raising, so it sat in the corner always ready to go. Since I wanted the quality cut of that heavy, high-quality machine, but couldn't afford the luxury, I decided to design and build my own shaper.

Before starting on my shaper, I thought it would be prudent to check out the available selection of stock raised-panel cutters. I discovered that there were relatively few straight-style raised-panel cutters available right off the shelf. I would need a different cutter for each job and carbide-tipped cutters are more than $100 each. So I quickly decided to take a different slant on shaper design to get a variety of profiles with a single cutter. This meant I needed a way to change the attitude of the cutter in relation to the table. Some very expensive shapers have tilting spindles for just such a purpose, but this system would have involved devising a mechanism to tilt the entire motor and spindle assembly—a complicated task. My more practical solution was to tilt the shaper table.

Building your own shaper, using a commercially made spindle assembly, is an economical way to add heavy-duty, panel-raising capacity to your shop. With its tilting table, welded-steel frame and 1-in.-dia. spindle, Perron's shop-built shaper shown here is tough enough to handle large-diameter panel-raising cutters and produce clean, chatter-free cuts.

My final design is shown in the photo below and figure 1 on the facing page. The shaper has a welded-steel base and employs a store-bought arbor assembly with a 1-in. spindle. The machine operates like a conventional shaper, except for the table: It tilts about 16° down and about 3° up. The machine weighs about 250 lbs., making it heavy enough not to dance all over the floor when I use it. Although the construction of the shaper involves quite a bit of metalworking (see "Metalworking in the Woodshop," *FWW* #79, p. 84), relatively few metalworking tools are needed to build the shaper, and you could substitute a wood base if you wish. Materials cost me about $350, including the spindle, but I used a lot of spare materials that were lying around my shop. While my machine was designed specifically for panel-raising, the basic shaper can handle any sort of cabinetmaking or millworking job, so you can customize it to suit your application. Whether or not you choose to duplicate the design in this article or alter it to your taste, I hope my experiences will inspire you to make your own shop-built shaper.

Building the base—Conventional shapers are built with the spindle/motor assembly hanging from the table, which in turn is mounted atop a base. Because I designed my table to tilt, the base became the heart of the machine, with motor and arbor bolted to it. The base, shown in figure 1, is made with welded steel tubing and angle iron, and has sheet metal panels on three of its sides. Foam insulation is glued to the inside surfaces of the panels to reduce noise. The top of the base is stepped, to provide clearance for the tilting table, and the 8-in. steel channel at the top of the welded base provides a mount for the shaper's spindle assembly. The two rear legs sport built-in casters that make the shaper easy for one person to move around.

All of the steel for my base came from a local scrap yard, so any suitable steel tubing or angle stock sizes you have lying around could be substituted for the ones I used. I cut all the pieces for the base frame to length with an abrasive blade in a motorized miter saw, and clamped the parts together. I used an arc welder and worked from the inside of the frame, so that the weld beads would not show, to give the machine a clean look. If you don't weld, you could have a friend or local welding shop make the base for you,

From *Fine Woodworking* (March 1990) 81:50-54

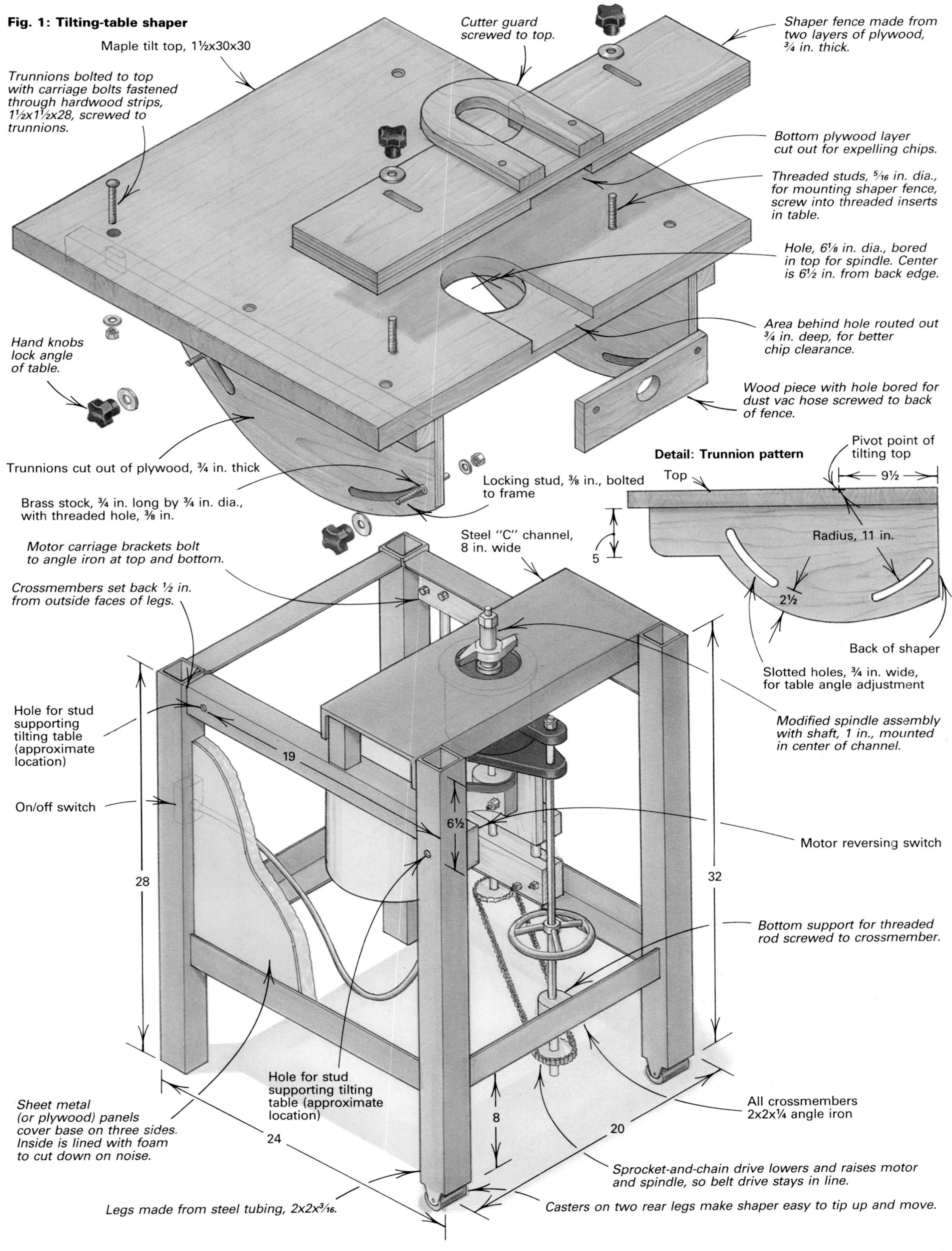

Drawings: Kathleen Rushton

Starting with the stock spindle he purchased, Perron modified the assembly to allow the cutter to be raised and lowered via a handwheel at the back of the shaper. A chain drive at the bottom links the handwheel shaft with a sliding motor carriage and adjusts the height of both the motor and spindle synchronously, to keep their V-belt drive pulleys aligned.

or you could make it out of wood. If you choose the latter alternative, use fairly large-dimension stock and employ strong, tight-fitting joinery, like mortise and tenon, so the base will be heavy and withstand vibration. After the frame was finished, I attached the metal side panels with pop rivets and bolted the casters to the rear legs, notched earlier. For anti-vibration padding and skid resistance, I glued pieces of neoprene, a synthetic rubber material, to the bottoms of both front legs.

Modifying the spindle—After unsuccessfully trying to design a safe and easy-to-build spindle that could adjust up and down and turn safely at 7,000 RPM, I decided to buy a commercially made spindle assembly. The one I selected is heavy and well made, with a 1-in.-dia. shaft (available from Mooradian Manufacturing Co., 1752 E. 23rd St., Los Angeles, Cal. 90058). I chose a 1-in. shaft because it fits the large panel cutters I wanted to use and because I wanted the shaper to stand up to heavy cutting pressures without deflecting. Although the stock spindle comes with a simple height adjustment screw, I wanted to be able to raise and lower the cutter without throwing off the alignment of the spindle and motor drive pulleys. Hence, I designed the height adjustment system, shown in figure 2 on the facing page, that consists of a modified spindle with a threaded rod running down to a sprocket and chain that in turn connects with a threaded rod that runs through a sliding motor carriage. When the handwheel on the spindle rod is turned, not only does the spindle (and hence cutter) height change, but the chain drive raises or lowers the motor as well, keeping the V-belt pulleys in line (see the photo above).

Modifying the stock spindle first requires drilling out the threads in the spindle housing flange. The lower flange on the moving part of the spindle is then fitted with a ½-13 Helicoil insert—a job I had done at a machine shop. Although you can thread the cast-iron flange, the Helicoil provides strong steel threads for the raising and lowering mechanism. One end of a ½-13, 24-in.-long threaded rod is locked to the upper flange with nuts that are drilled through so they can be secured to the rod with cotter pins. Next, the spindle assembly is bolted to the underside of the channel atop the base, with a 1⁄32-in. neoprene washer between them to act as a gasket. I scrounged a handwheel from an old lathe to use as the adjustment wheel, and I locked this on the threaded rod with a pin. A stop nut, to limit the travel of the spindle to about 3 in., is positioned and cotter-pinned to the rod, as shown in figure 2. The bottom end of the threaded rod is held by a homemade pillow block—a block of wood drilled to fit the rod and screwed to the angle iron crossmember at the back of the base.

At this stage, I tested the modified spindle and it worked well, except that the fit between the inner spindle sleeve and the outer housing was so tight that sliding them was difficult. To remedy this, I took the assembly apart, cleaned the surfaces and coated them with Molykote 321-R, a dry lubricant designed for use with tightly mating parts (made by Dow Corning and available from your local bearing supply store). After reassembly, the spindle adjusted much more easily. I fitted my shaper with a special spindle lock (you can see the knob beneath the table in the photo on p. 104), but the stock spindle lock is also easy to use.

Making the sliding motor carriage—The motor carriage acts as a mount for the shaper's motor and allows the motor to travel up and down in unison with the spindle. The carriage is made from four hardwood scraps, a square of ¾-in. plywood, two 20-in. lengths of ½-in. galvanized pipe (that act as guide rails) and a 22-in.-long, ½-in. threaded rod raising screw. For the carriage to slide smoothly, the holes for the pipe and rod need to be drilled accurately. Therefore, it's best to line up and clamp together the four hardwood pieces (dimensioned as shown in figure 2) and then drill them all at once. I ground down a ⅞-in. spade bit until it was a bit larger than the OD of ½-in. pipe. Also, drill a slightly oversize hole for the ½-in. threaded rod. Two of the hardwood pieces become the pipe holders and receive bandsawn kerfs and holes for ¼-in. carriage bolts in both ends. The other two pieces are glued and screwed to the plywood square, which has holes for mounting the motor. A piece of ¼-in.-thick brass plate, tapped to fit the threaded rod, is screwed to the motor platform as shown. After the pipes are slipped through the holes in the platform, which may require reaming so the platform slides smoothly over the pipe, they are fitted in their holders and the carriage bolts are tightened to lock the pipes in place. The threaded rod is then screwed into its plate and nuts and washers are fitted at each end. The nut at the top is recessed in a counterbored hole, and both nuts are pinned through the rod, to keep them from moving.

Once the carriage is completed, the motor is bolted on its platform and the entire assembly is bolted inside the base. My motor is a 1-HP, 3,450-RPM capacitor-start motor. While this has proved adequate in most conditions, you may opt for something a little more powerful, say a 2-HP motor. I mounted a metal on/off switch box to the front of the base, and also a reversing switch, fixed on the back of the machine to eliminate someone accidentally mixing it up with the on/off switch. A reversing switch is often necessary on a shaper, so that cutting direction can be changed. A V-belt connects the motor and spindle pulleys. The pulley ratio is a little more than 2-to-1, which yields a spindle speed of about 7,000 RPM. I

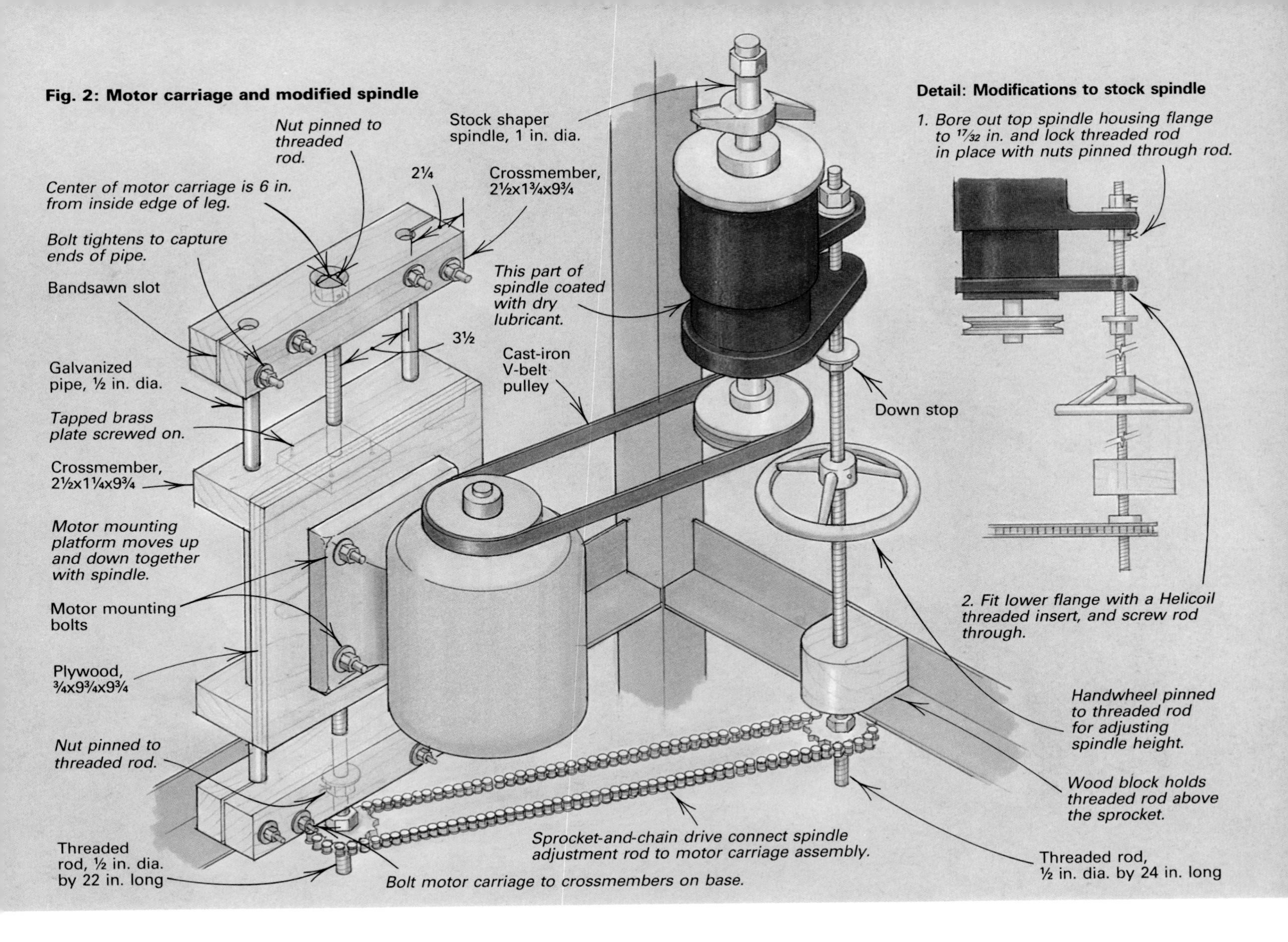

chose Browning cast-iron pulleys fitted with split-taper bushings that grab the shaft they're mounted on like a collet. At $30 apiece, these pulleys aren't cheap, but they're precise and won't come loose during high-speed operation. The bottom ends of the threaded rods on both the motor carriage and spindle are fitted with 2-in.-dia. sprockets, screwed on and pinned in place, and tied together with no. 35 roller chain. Sprockets, chain and pulleys are available from your local motor and bearing supply house.

The tilting table—Before I could build the tilting mechanism for my shaper, I had to determine the best pivot point: an axis for the top to turn around. Because I had decided to mainly use my shaper with 5¾-in.-dia. raised-panel cutters, I set the pivot point at the intersection of the tabletop and the outer shoulder of the cutter (see the detail in figure 1 on p. 105). Pivoting at this point allows me to set the height of the shoulder of the cut to stay the same depth, regardless of the angle of the table. Tilting allows me to get the desired thickness I want on the outer edge of a panel, regardless of the thickness of the stock I'm using.

The pivot mechanism I designed is a trunnion-type system, similar to the mechanism used on many bandsaws and sanders to tilt their tables. Two plywood trunnions attach to the underside of the table, each with a pair of curved slots that ride on studs, which are 4-in. lengths of ⅜-in. threaded rod, bolted to the base. A hand knob on each stud clamps the trunnion tightly against the base, locking the table in place. A bushing that rides in the trunnion slot is made from a ¾-in.-long, ¾-in.-dia. piece of brass stock, center-bored and tapped to screw on the stud.

Trunnions are bandsawn from a good grade of birch plywood, with no voids in the core layers, following the pattern shown in figure 1 on p. 105. The curved slots are cut with a ¾-in. straight bit in a router mounted on a swing-arm jig. The radius of these is 11 in. to the center of the slot, as measured from the desired pivot point of the table. To make the slots, the swing arm's pivot point is set in a small scrap of plywood, temporarily tacked to the top edge of the trunnion.

The shaper's 1½x30x30 tabletop is glued up from strips of 8/4 maple. Since my thickness planer only handles 12-in.-wide stock, I glued up the top in three 10-in.-wide sections. After planing each section flat, I glued the three panels together. To me, unfinished maple is smooth enough for the table, but you can make the top out of plywood or particleboard and cover it with plastic laminate, for a slicker, more wear-resistant surface. Then, a fly-cutter chucked in the drill press was used to cut out the 6⅛-in.-dia. hole for the cutter. Later, I decided to rout a ¾-in.-deep recess at the back of the cutter opening to increase the shaper's chip clearance.

The top is fastened to the plywood trunnions with carriage bolts through two 1½x1¾x28 hardwood strips screwed to the trunnions. These strips stiffen the top and provide a strong joint between it and the trunnions. Two slotted bolt holes at the ends of each strip allow for expansion and contraction of the solid-wood top.

Once the trunnion sections are attached to the top, the whole assembly is ready to be mounted on the base. First, the surface of the table is positioned square to the spindle in all directions. I use

By tilting the shaper table relative to the cutter, Perron gets a variety of raised-panel profiles from a single cutter—a real bonus since carbide cutters are expensive. The U-shape guard on the adjustable fence protects Perron's hands as he feeds a cherry panel by the cutter.

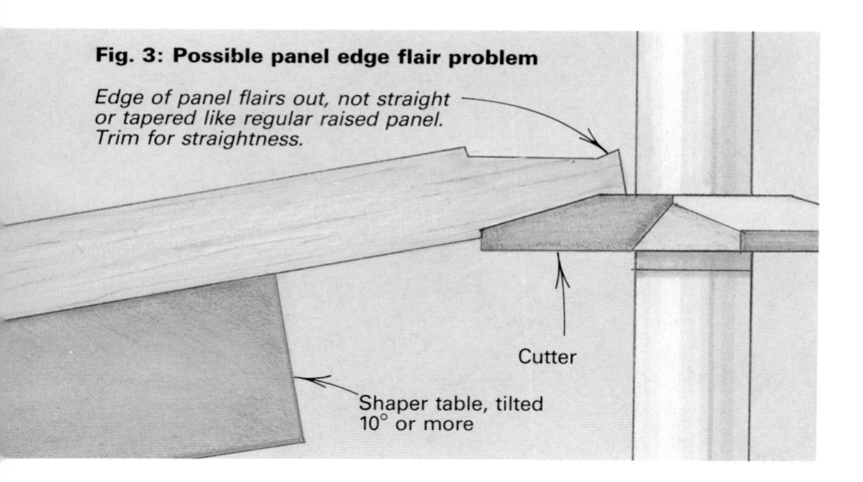

wedges and shims made from scraps to align the table, and then use a pair of pipe clamps attached across the trunnions to temporarily lock the table in place. Next, the position of the locking studs is marked on the base, using the slots as a guide. With a metal punch, I mark the center position of each stud, one on the top crossmember and one on the rear leg, for each trunnion. After removing the table and drilling the ⅜-in. holes, the studs are installed with nuts and lock washers. The brass bushings must be shimmed out slightly on the crossmember studs to make them flush with the rear leg studs. Large washers fit under the clamping knobs, to distribute the clamping pressure on the plywood trunnions. I purchased easy-to-grip locking knobs from Reid Tool Supply Co., 2265 Black Creek Road, Muskegon, Mich. 49444.

The fence—Even on a commercially made shaper, I'll sometimes set the stock fence aside and make a custom one for a specific job or cutter setup. By doing this, I can make the operation safer and improve chip clearance. I made my shaper fence from laminated plywood topped with a guard that overhangs the cutter to keep hands out of harm's way. With the protection this fence affords, I feel safe raising panels as narrow as 10 in. wide, holding them free-hand. The fence is notched, to clear the top of the spindle, and the lower part is cut away completely behind the cutter, for chip exhaust. The outside profile of the U-shape guard is bandsawn slightly larger in diameter than the cutter and attached to the top of the fence with two drywall screws. Two 3-in.-long, ⅜-in. studs that fit through slotted holes cut in the fence attach the fence to the tabletop. The studs are screwed into threaded inserts installed in the tabletop. To take care of the prolific amount of dust and chips that panel-raising produces, I bored a hole in a wood scrap to fit the hose on my dust-vac and screwed this to the back of the fence (see the photo on p. 106).

My panel raiser is a joy to use, it's smooth running and I like listening to it hum. It is satisfying to run a piece of stock by the cutter and flip it over to find a panel bevel that needs no sanding. The only problem I have had is with panels cut using the full width of the panel cutter with the table tilted to 10° or more (see figure 3 at left). The panel edges flair slightly, preventing proper seating in a frame groove or rabbet. However, the problem can be eliminated by not using the full width of the cutter when raising, or by trimming the flared edge with a straight cutter in the shaper or with a handplane. □

Frank Perron is a Field Experiment Coordinator for the U.S. Army Corps of Engineers and part-time woodworker living in Pomfret, Vt.

Safety warning

The shaper has the potential to be a dangerous machine. The rotational force of a large panel-raising cutter can grab even a large panel out of your hands and hurl it with surprising force. Worse, the force can unexpectedly pull your hand into the cutter, with disastrous results. Therefore, *always use the shaper with the guard in place and keep your hands well clear of the cutter*, wear safety glasses or a face shield and avoid wearing jewelry or loose clothing. As with any home-built machine, make sure all parts are properly installed, tightened and adjusted before use. Also when raising panels, always take several shallow passes rather than one deep cut.

A cheap saw kit the author encountered at a 1950s industrial-arts show inspired the design for this shopmade parallel-arm scroll saw. Powered by an old dryer motor, the saw's 2-in. stroke makes for an aggressive cut.

Shopmade Scroll Saw

Eccentric drive simplifies construction

by Mark White

I was a high-school student in the mid-1950s when I attended my first industrial-arts conference in Oswego, N.Y. One of the displays there featured a cheap kit for a motorized version of the wooden scroll saw my great uncles had used to cut fretwork for fancy houses they built at the turn of the century. The saw was made almost entirely of unfinished, ¾-in. ash. It was simple and homely, but boy could it cut. For years, I've tried to lay my hands on one, but the manufacturer has disappeared without a trace.

Since I had already built a large walking-beam saw (see *FWW #24* or *FWW on Making and Modifying Machines*), I finally decided to design my own scroll saw. I ended up with a saw that performs as well as any of the factory-made machines I've tried and will saw very tight curves in wood up to 2 in. thick, leaving a smooth surface that requires no sanding. For inside cuts on fretwork, the blade can be removed, threaded through a hole in the wood and reinstalled in less than 30 seconds.

Inspired by the homely kit I'd seen years earlier, I made my saw as simple as possible. Basically, it consists of two parallel wooden arms mounted on a rigid wooden frame and kept in tension by the blade at one end and a stout nylon cord at the other end. The blade is driven by the reciprocating motion generated by a pair of eccentric, rotating weights attached to the lower arm with a shaft and pillow block. An old clothes-dryer motor drives a section of rubber hose that acts as a flexible shaft to spin the weights. Because the weights are eccentrically mounted, they actually unbalance the pillow-block shaft, causing it to oscillate one cycle for every revolution of the motor. Although the stroke can be varied by changing the length of the weights, I've found that weights made from 2-in. bar stock, 3½ in. long, work best with the 6-in. coping sawblades I use in the saw. These longer blades produce a more aggressive cut than the 4-in. blades most scroll saws use. As shown here, the saw has a 24-in. throat, but if you need a smaller or larger throat, you can scale the dimensions up or down without affecting performance.

Building the frame and arms—When choosing wood for the frame, pick a stable and warp-resistant material. For this saw, I used two Sitka-spruce 2x12s about 32 in. long, because that's what I had. White pine or fir would work as well, or if you want a nicer-looking saw, use walnut, beech or maple. To ensure precise alignment of the arm pivot holes, screw the boards for the frame together and machine both parts at once. The pivot holes—½ in. dia. for the top arm and ¾ in. dia. for the bottom arm—should be bored on a drill press. Bore the large access holes for the rotating weights with a Forstner bit or a hole saw, or saw them out with a jigsaw and much patience. After bandsawing the frames to shape, I separate the two sections and rout a radius on all the edges except those that mate to the base and the saw table.

I also used spruce for the arms, but ash, cherry and pine are good choices, too. Whatever wood you pick, keep the weight of the arms as low as possible near the blade end. If the arms are too heavy, excess vibration and poor reciprocating action will result. One advantage of spruce is that it has sound knots hard enough to serve as bearings for the bolts or rods on which each arm pivots. On one of the saws I've built, a 1¼-in. knot drilled for the pivot shaft and regularly lubricated with gear oil has lasted four years as a bearing. The lower pivot on the saw shown here is a ¾-in.-dia. steel rod in an Oilite bearing of the equivalent inside diameter. Oilite bearings—self-lubricating, sintered bronze sleeves—are available in a range of sizes. Refer to the supplies box for a source.

As the drawing shows, the distance from the pivot point to the blade mount is ¼ in. longer on the top arm than it is on the

Photos: Marion Stirrup

bottom arm. This causes the blade to back out of the cut on the upstroke and advance into the material on the downstroke, increasing blade life and improving performance. Reducing the blade rake (angle) to ⅛ in. will make the blade less aggressive and the cut more precise.

The eccentric drive—For the drive, you'll need some ¼-in. by 2-in. mild steel bar stock, a 3½-in. to 3¾-in. bolt to serve as a shaft and a ¾-in. ID pillow block. As shown in detail C, the weights are fashioned from the bar stock, then mounted on either side of the pillow-block bearing by the shaft. The threads on the bolt's excess length are filed or ground off to give the rubber-hose coupler good purchase. To make the weights, I first bored the bar stock, then cut them to final size. It's tough boring large holes in heavy steel, so I started by boring a ¼-in. hole, which I then enlarged to ¾ in. Do this boring on a drill press and be sure to clamp the steel firmly and keep your hands well clear of the work.

I assembled the drive mechanism by holding the bolt head tightly in a vise while the nut was drawn up very lightly. Washers, or a ⅞-in. nut, can be used as spacers to keep the weights from striking the arm as they rotate. As you tighten the nut, make sure the weights are aligned so they'll rotate in unison, otherwise the reciprocating action will be uneven. Once it's assembled, position the pillow block on the lower arm, as shown in the drawing. Before bolting the pillow block down, make sure the weights rotate through their full arc without striking the arm. If they strike the arm, add a thicker spacer. By the way, the drive can be positioned so the motor is on either side of the saw.

Assembling the saw—Begin assembly by inserting the arm pivot bolts in their holes and positioning the upper and lower arms on the frame. Rotate the eccentric weights by hand, and on the inside of the frame, mark the path they describe. With a large Forstner bit and/or a chisel, chop clearance cavities in both frame pieces to accommodate the rotating weights. Remember, the arm's travel is at least 1 in. in both directions, so be sure you've provided enough clearance. Although the spinning weights are well protected by the saw's frames, it's probably not a bad idea to fashion some sort of a removable guard for the back of the saw as an added safety feature. Before proceeding with final assembly, the blade holders and tensioner must be made.

I made the blade holders from ⅜-in. key stock, as shown in the drawing. Each blade holder fits into a slot cut into the end of the arm and is held in place by a $^3/_{16}$-in. steel pin. If you can find them, pins that are hardened and ground will work best with the softer key stock, but in a pinch, a small bolt could also suffice. The blade itself is inserted through the holder's slot and held in place by pins on the ends of the blade.

To cut well, the blade must be under considerable tension; and on commercial saws, this is usually done with a threaded rod. But in keeping with my saw's low-tech design, my tensioner is simply a loop of nylon cord that passes around a ¼-in. pin in the lower arm and through a hole in the upper arm, where it wraps around a dowel. Twisting the loop tensions the blade, as with a bowsaw. This setup may sound crude, but it's effective, and because it's flexible, the saw won't shake itself apart when a blade breaks. Sometimes, vibration will tend to unwind the tensioner, a problem that can be remedied by carving a detent notch for the dowel where it seats against its mounting knob.

To minimize warping, I finish both sides of the frames with varnish or shellac before assembling the saw. Once the finish is dry, I test assemble the parts, tightening the fasteners fingertight.

Fig. 1: Scroll saw plan

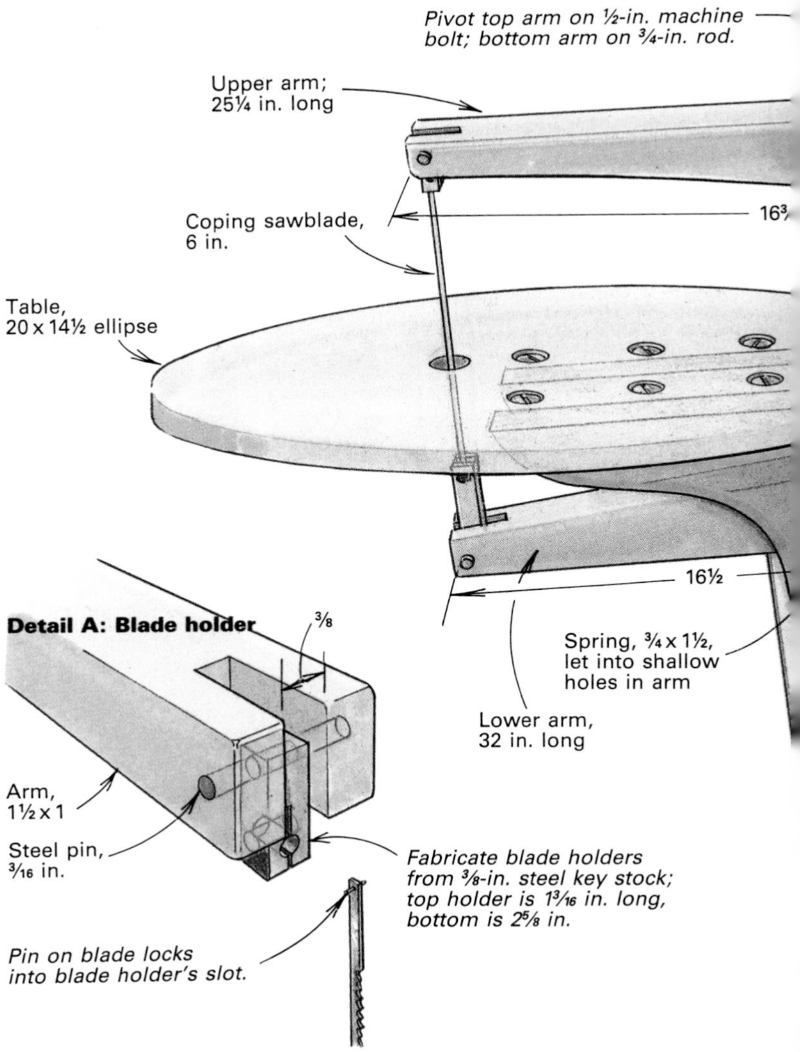

Pivot top arm on ½-in. machine bolt; bottom arm on ¾-in. rod.

Upper arm; 25¼ in. long

Coping sawblade, 6 in.

Table, 20 x 14½ ellipse

16½

Spring, ¾ x 1½, let into shallow holes in arm

Lower arm, 32 in. long

Detail A: Blade holder

⅜

Arm, 1½ x 1

Steel pin, $^3/_{16}$ in.

Fabricate blade holders from ⅜-in. steel key stock; top holder is $1^{3}/_{16}$ in. long, bottom is 2⅝ in.

Pin on blade locks into blade holder's slot.

With a blade installed and lightly tensioned, I move the arms by hand. They should slide lightly against the sides of the frame. If there's binding, trim as needed with a handplane. To keep the lower arm roughly centered in its swing and to give the rotating weights some resistance to work against, I mounted three coil springs between the lower arm and the saw's base. The springs—straight from the hardware store—are 1½ in. long, ¾ in. in diameter. To hold each spring fast against vibration, I bent one end of the coil down and threaded it into a small hole bored in the base. Long finishing nails will temporarily hold the coil springs in place while the saw is attached to its plywood base.

Trial run—To test the saw, I chuck a bolt into a variable-speed drill and connect this through a section of rubber hose to the eccentric drive shaft. I run the machine for a few minutes at slow speed to check everything out. Both arms should reciprocate freely with minimal vibration. If the front or back of the bottom arm strikes the base, adjust the position of the springs or install stiffer ones. Once this test is done, I connect a permanent motor and switch. The saw doesn't require much power—⅓ HP to ¼ HP should be plenty at 1,720 RPM. Do not, under any circumstances, use a 3,450-RPM motor. Unless you reduce its speed through pulleys, a motor this fast will cause dangerous vibration.

To finish up the saw, I make a 20-in.-dia. elliptical saw table out of ¾-in. plywood and screw it to the frames with drywall

From *Fine Woodworking* (May 1988) 70:51-53

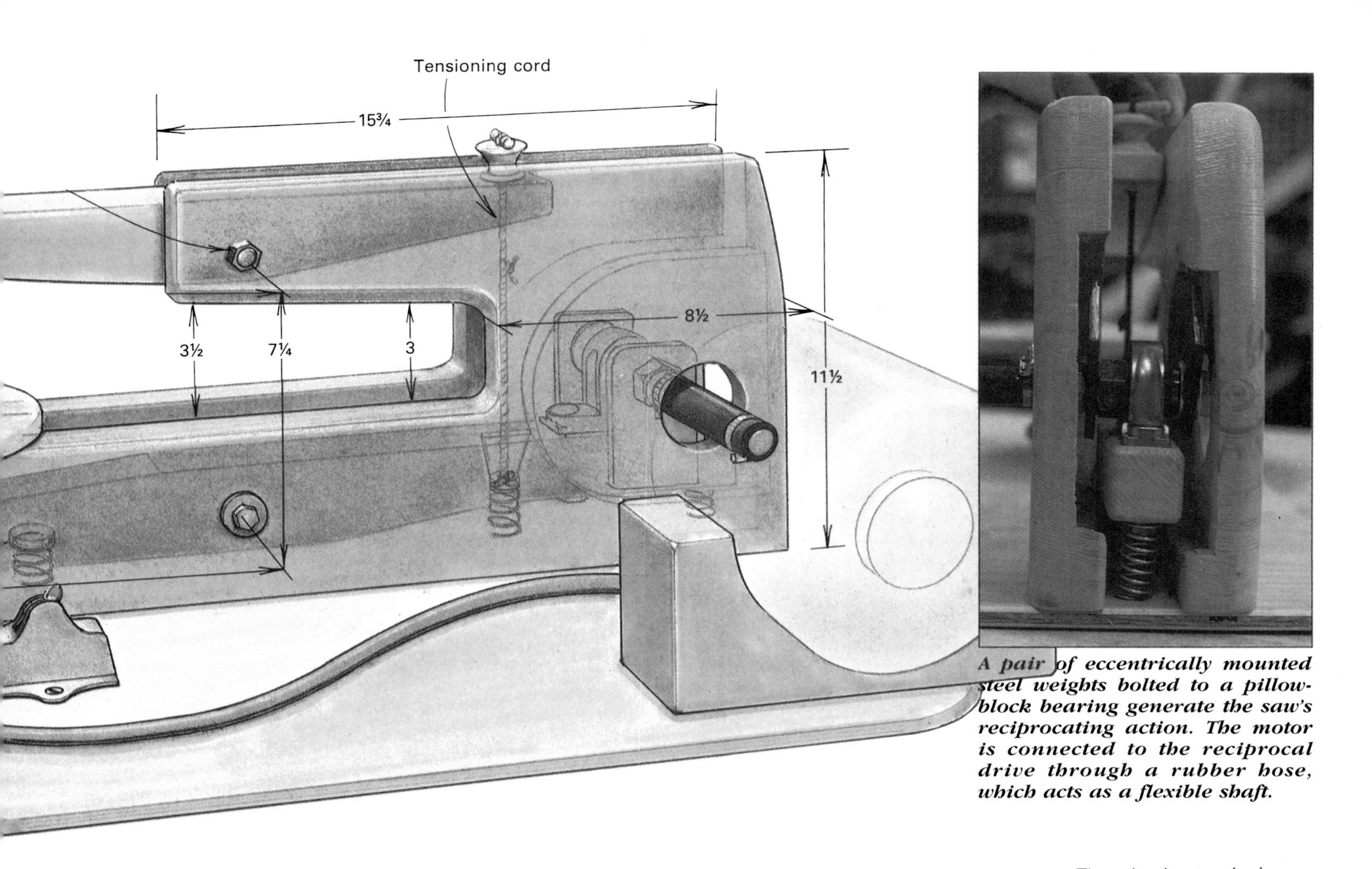

A pair of eccentrically mounted steel weights bolted to a pillow-block bearing generate the saw's reciprocating action. The motor is connected to the reciprocal drive through a rubber hose, which acts as a flexible shaft.

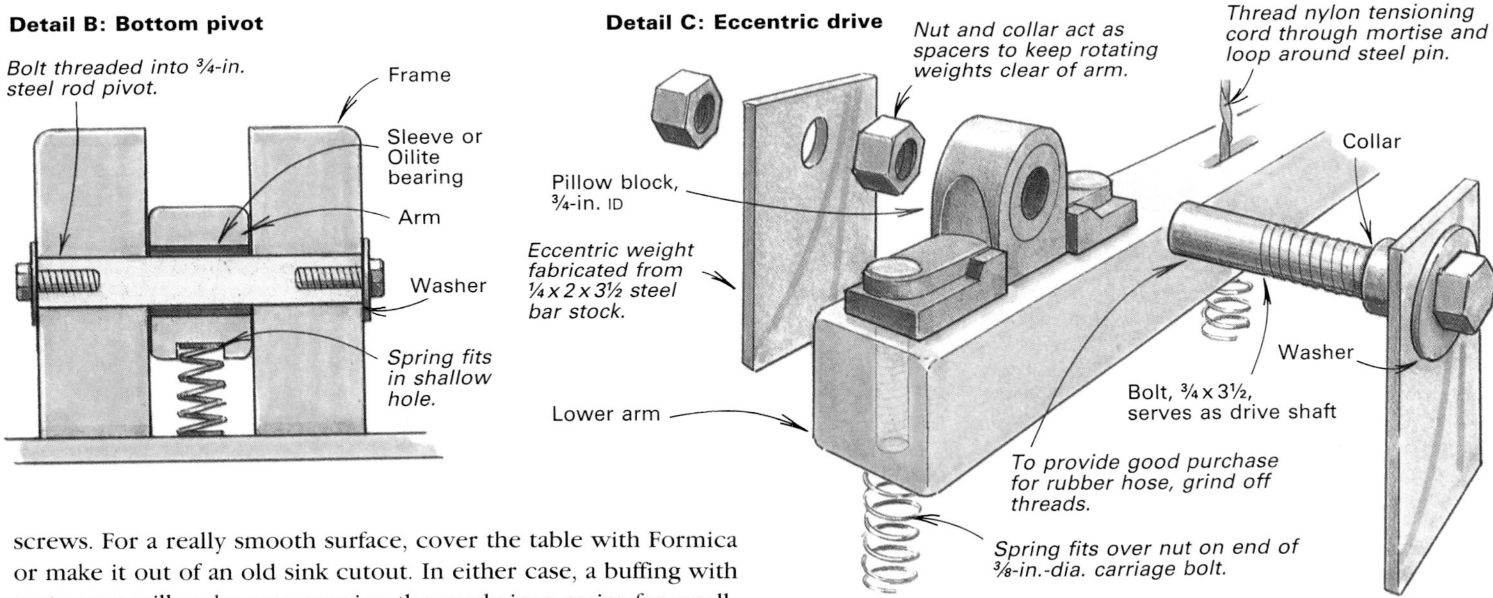

screws. For a really smooth surface, cover the table with Formica or make it out of an old sink cutout. In either case, a buffing with paste wax will make maneuvering the workpiece easier for small-radius scroll work. I've come up with solutions for the two aspects of using a scroll saw that I find most unpleasant: vibration and dust. A 3-in.-thick foam-rubber pad placed under the saw's base dampens noise and vibration considerably and keeps the saw from walking across the table. On a few of the machines I've built, I tapped into the airstream coming off the motor's cooling fan and diverted it through a ½-in. copper tube to a point just in front of the blade. If you do this, make sure to orient the tube so it blows dust toward the back of the saw and not toward the operator. □

Mark White teaches woodworking, welding and house construction at the University of Alaska outpost on Kodiak Island.

Sources of supply

Pillow blocks and motors are available locally from Grainger's. For a catalog and list of distributors, write W.W. Grainger, Inc., 5959 W. Howard St., Niles, IL 60648; (312) 647-8900.

Bar stock, Oilite bearings and hardware are available from Small Parts, 6901 N.E. Third Ave., Miami, FL 33238; (305) 751-0856.

Key and bar stock is available from Metal by Mail, 18170 W. Davidson Road, Brookfield, WI 53005; (414) 786-4276.

Building a Stationary Sander

by Roger Heitzman

Although I've built more than half a dozen woodworking machines from scratch, it's often desirable to take an easier course and start with premachined parts or kits. I was in a friend's woodshop one day and noticed an old stationary belt sander, which he had built from a Gilliom kit, gathering dust in the corner because he had replaced it with a newer, larger-capacity sander. After negotiating the sale, I took the sander home, figured out how I could modify its basic design to suit my needs and built the sander shown in the photo on the facing page. The Gilliom was designed to pivot in one axis, to sand vertically or horizontally, but what I really needed was an edge sander. Therefore, I designed a mechanism to allow the entire sander and motor to pivot the belt along its long axis for edge sanding the long edges of a panel—a very convenient operation. The four main components of the tilt mechanism, shown in figure 1 below, are machined from aluminum plate, drilled, tapped and assembled using techniques outlined in "Metalworking in the Woodshop" (*FWW* #79, p. 84). The motor and sander bolt to the motor bracket that attaches to the pivot bracket through a slotted plate, which holds the main pivot shaft and tilt-position locking stud. The slotted holes in the plate allow the entire motor/sander to be shifted from side to side so its weight can be adjusted for balance, which makes pivoting easier. The pivot bracket bolts the entire assembly to the stand. All the motor mounting and fastener holes and slotted holes are countersunk so the Allen bolt heads don't hang up when the mechanism is pivoted.

I built the stand with 2-in. by 4-in. steel tubing braced with a smaller-size square tube. After cutting the parts to length with an abrasive wheel in the chop box, I arc-welded the stand together. I also welded several ½-in.-dia. steel rods on one side of the horizontal brace as hangers for belt storage. In addition, I added casters for mobility, and two retractable feet to stabilize the sander once it's in position.

The original kit sander only has one table, but with my expanded edge-sanding capacity, I needed two adjustable tables, as shown in figure 2 on the facing page. I made both the vertical sanding and edge sanding tables from ½-in.-thick micarta. I bolted the edge sanding table to an adjustable support that provides side-to-side movement for positioning the table when bevel sanding and up-and-down motion for sanding on different areas of the belt. I

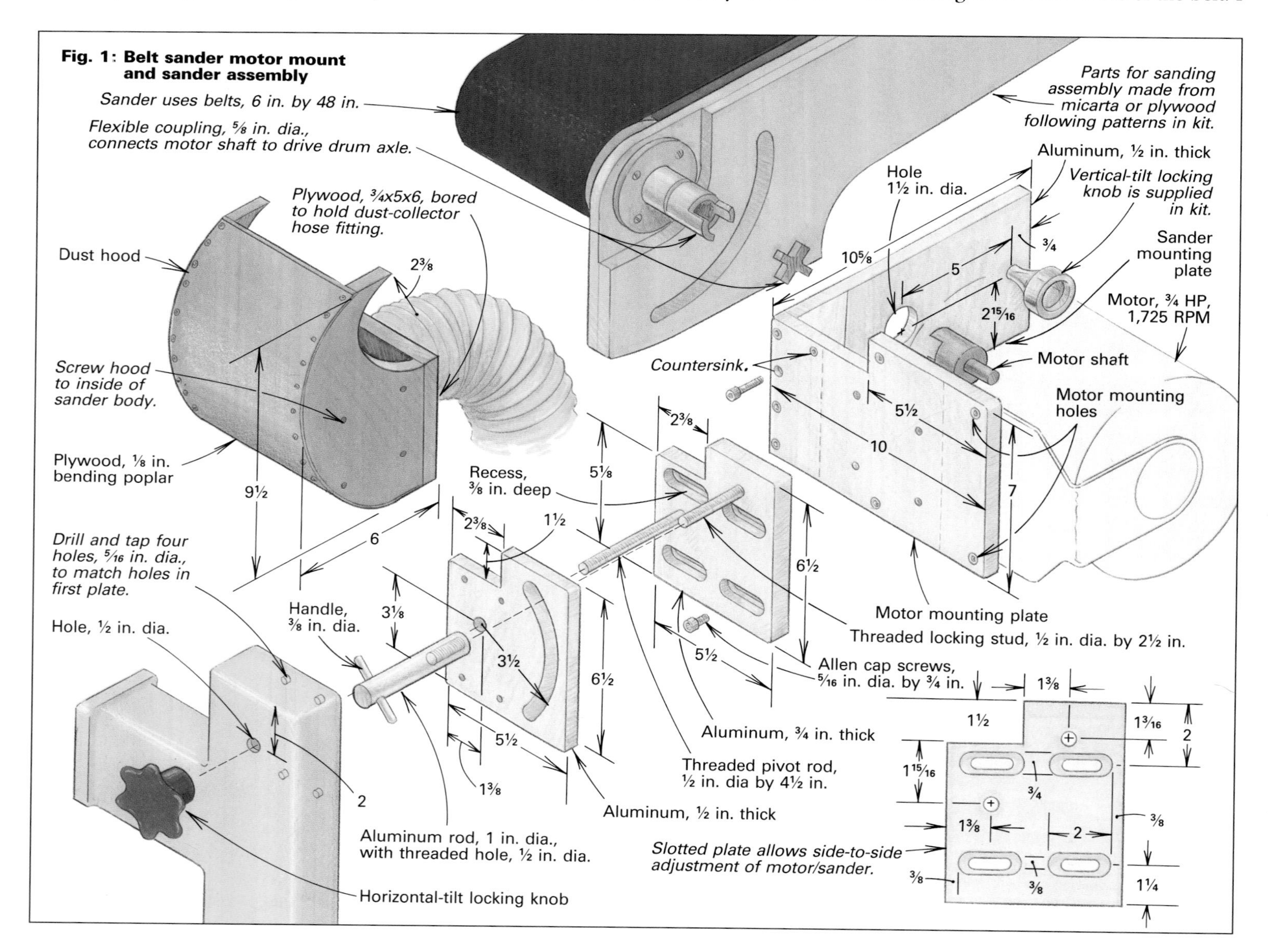

From *Fine Woodworking* (November 1989) 79:88-89

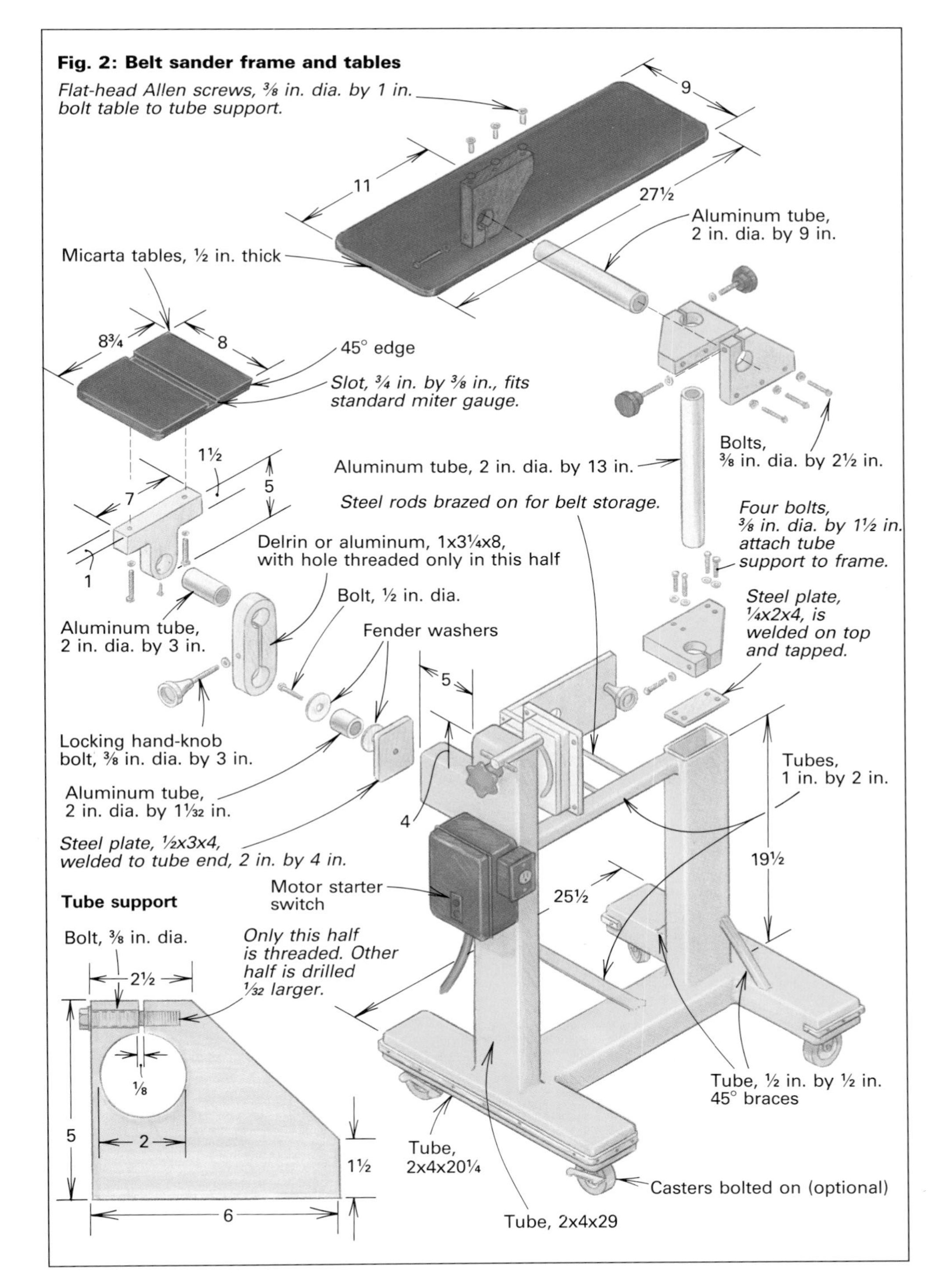

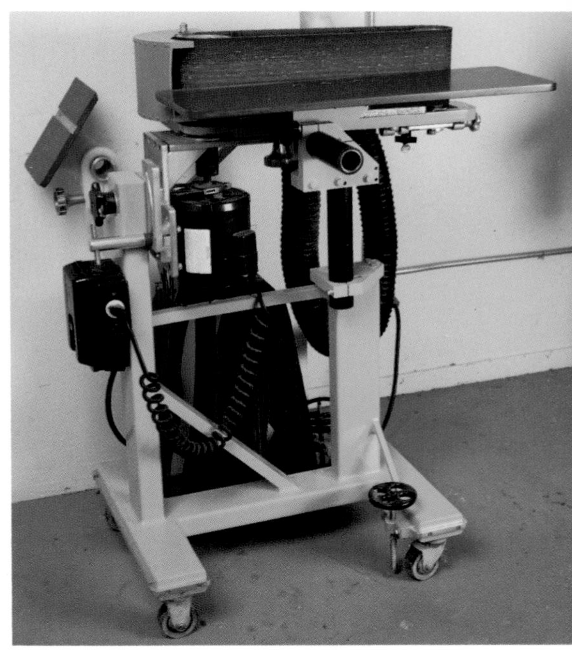

Starting with parts from a Gilliom belt sander kit, the author welded a steel stand and machined aluminum and plastic parts to create a customized sander. Both a special tilting mechanism and direct motor drive allow the sander to do horizontal, vertical and edge sanding.

built the table supports from heavy, 2-in. OD aluminum tubing and 1-in.-thick mounting blocks I found at a local surplus yard. The vertical sanding table is articulated on an oval-shape arm, machined from delrin (aluminum would also be good) that attaches and adjusts via short sections of aluminum tubing. The vertical table has a slot machined in it to fit a standard saw miter gauge. Except for the two lock knobs I scrounged from the Gilliom kit, I used threaded hand knobs, available from MSC Industrial Supply Co., 151 Sunnyside Blvd., Plainview, N.Y. 11803; (800) 645-7270, to lock all the adjustments on both tables. I elected to do without an angle scale and pointer for either table: I set angled sanding operations with a bevel gauge.

The 6-in. belt sander itself is built pretty much as it comes from the Gilliom kit, except you don't need to make any of their table or stand parts (the kit is available for $99.50 from Gilliom Manufacturing Inc., Box 1018, St. Charles, Mo. 63302; 314-724-1812). When I rebuilt the sander I bought from my friend, I machined most of the new parts from micarta because it's stable and durable. The exception is the belt platen, which I made by laminating particleboard to a 1¾-in. thickness as the kit recommends. To keep the belt running effortlessly, I stapled graphite cloth (available from Derda Inc., 1195 W. Bertrand Rd., Niles, Mich. 49120; 616-683-6666) over the top of the platen. The assembled sander bolts to my homemade L-bracket I described earlier, via the kit-supplied spindle-mounting flange.

I used a ¾-HP motor to power the sander, switched on and off by a mechanical electrical motor starter that's bolted to the stand. A magnetic starter would be better, but it's more expensive. The standard Gilliom kit uses a belt-and-pulley drive setup, which I converted to direct drive via a flexible drive coupling, available from W.W. Grainger Inc., 5959 W. Howard St., Niles, Ill. 60648; (312) 647-8900. Call or write for the address of the regional distribution center nearest you. To control the sander's copious dust production, I made a dust hood from scrap plywood with a hole for a vacuum-hose fitting, and bolted this to the inside surface of the sander's carriage panel, as shown in figure 1 on the facing page.

I tested the completed sander and was so pleased with how well it worked I decided to give it a deluxe finish. I first used body putty to smooth over the rough welded seams and fillets. After sanding the entire frame smooth, I cleaned all the metal surfaces with mineral spirits, and then I sprayed the frame with one coat of metal primer and two final coats of yellow gloss enamel. My stationary belt sander is definitely the nicest looking machine in my shop, which in part accounts for why it's one of my favorites. □

Roger Heitzman teaches an annual seminar on building machinery through the Baulines Crafts Guild. For more information, call the Guild at (415) 331-8520.

Drawings: Roland Wolf

Narrow-Belt Strip Sander

Shop-built workhorse for shaping, sharpening and smoothing

by Robert M. Vaughan

My narrow-belt strip sander is one of the handiest tools in my shop. It's great for easing or beveling edges, rounding corners, sharpening dowels, fudging miters for a perfect fit and smoothing the bandsawn edges of straight and curved surfaces. In addition to working wood safely and precisely, it's a metalworking tool, perfect for sharpening lathe tools and drill bits, deburring rough edges and shaping metal parts.

I built the sander shown below, because most of the $100-and-up store-bought models vibrated excessively and were too flimsy to be accurate. I wanted a sander that would be stable without being bolted to a bench, inexpensive and easy to build with normal workshop tools and readily available hardware (ideally, the odds and ends hanging around my shop).

My sander is nearly the same size and weight (about 48 lbs.) as many commercial models, but my hardwood-and-plywood frame absorbs vibration much more effectively than the plastic and sheet-metal commercial units. I used oak and ash for the frame on the sander shown below, but because these ring-porous woods deflect drill bits, they make it difficult to properly center a hole. I'd recommend using maple for the frame members. The frame supports a large, Formica-laminated work surface that can be tilted through 45°. Although the sander uses 1-in.- wide, 42-in.-long belts, available from Sears and several other companies, you can also use 1-in. strips ripped from 6-in. by 48-in. belts. The belt runs over three wood wheels. One of the two idler wheels attaches to a spring-loaded, pivoting upper arm, which tensions the belt. Belt tracking is controlled by adjusting this idler wheel with two counteracting thumbscrews.

Vaughan's narrow-belt sander, assembled from scraps of oak, plywood, fiberboard and assorted hardware, is stable, accurate and capable of performing a multitude of tasks.

A salvaged ⅓-HP, 1,750-RPM motor turns a 3½-in. pulley, which is V-belted to a 3-in. pulley on the mandrel. With a 3⅝-in. drive wheel, the belt moves at a rate of 1,900 surface feet per minute (SFM). Commercially available sanders usually run 3,000 SFM to 6,000 SFM and vibrate a whole lot more than mine does.

Figure 1 on the facing page shows the dimensions of the hardware and fittings on my sander. Your hardware may be different, especially if you scrounge your materials as I did, so collect the metal parts before you begin construction, and adjust the dimensions as needed. The mandrel is the heart of the sander, so look for one with ball bearings, a threaded shaft on one end for the drive wheel and a plain shaft on the other end for the pulley. Even if you follow the plan exactly, you'll want to trial-fit and fine-tune most of the parts before assembly. I clamped the parts together and did not drill any pilot holes or install any screws until I was sure of the fit.

Building the base and frame—There are six basic wood parts in the frame, including the ¾-in.-thick plywood baseplate on which everything is mounted. The other components, all cut from 8/4 hardwood, are the base block supporting the mandrel, the trunnion block holding the table, the back post, the top arm and the platen block, which supports the sanding surface.

After cutting the plywood base, I run countersunk screws up through the plywood into the base block. The base block must be wide enough to support the drive-shaft mandrel and about 10½ in. long, as shown in figure 1. You will have to remove this base block later, so don't glue it down. Set the mandrel on the base block with its centerline about 3¾ in. back from the front edge; screw it down with the largest wood screws that fit the mandrel base holes. You can't accurately align the mandrel without the drive wheel, so the next step is to build that wheel and the other two for the sander.

Wheels, shafts and bearings—The wheels are seven pieces of ¼-in.-thick tempered hardboard laminated together with Titebond glue. Hardboard, available from most large building-supply houses, has no voids or dense spots that could cause vibration or balance problems. The drive wheel, bored ⅝ in. for the mandrel's drive shaft, is 3⅝ in. in diameter, while the two idler wheels are 3¼ in. in diameter and bored out to accept two stan-

From *Fine Woodworking* (March 1989) 75:62-65

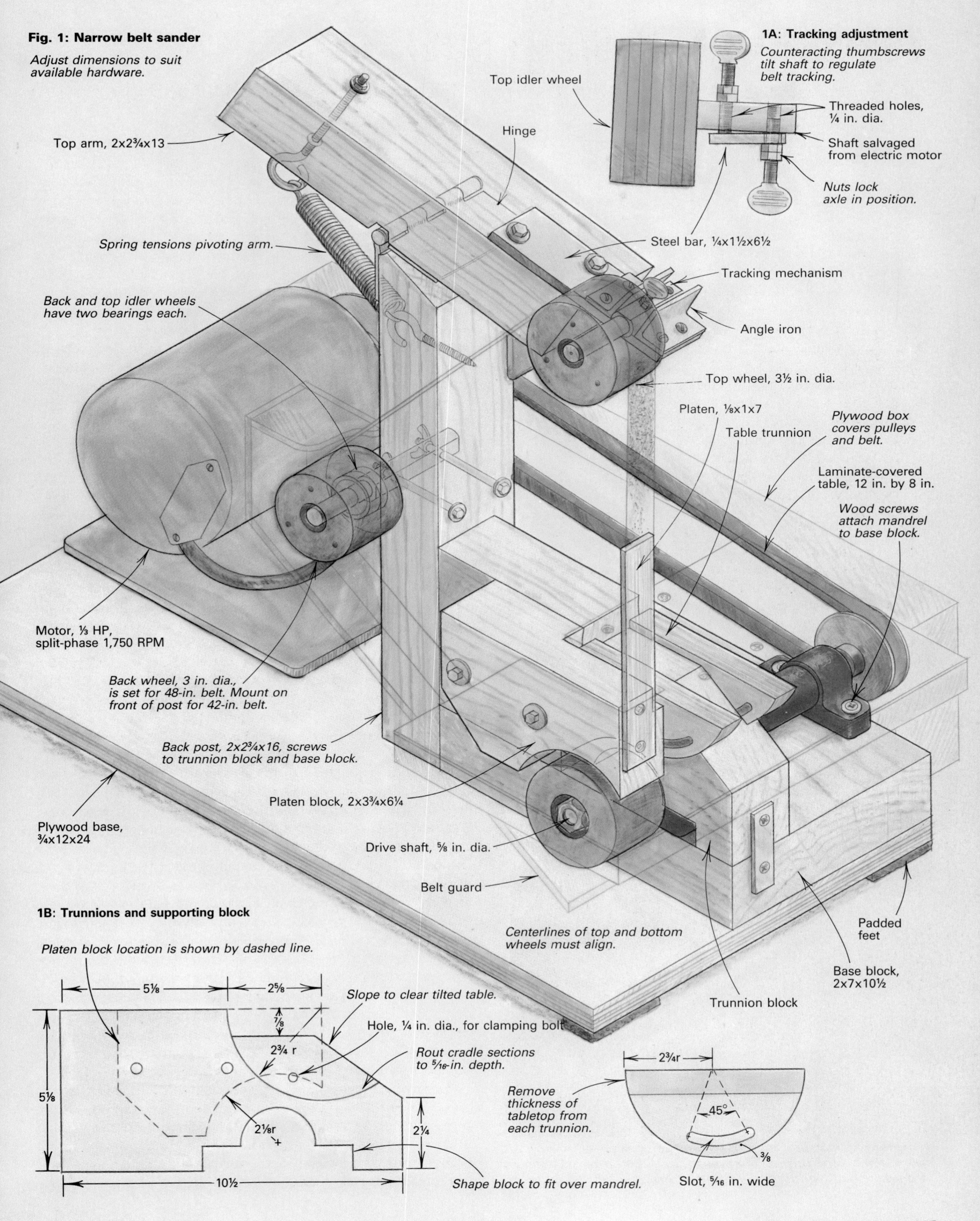

Drawing: Roland Wolf

dard 40mm-dia. by 12mm-thick ball bearings. I bandsaw each wheel slightly oversize, mount it on a lathe faceplate and turn its outside diameter with a slight crown—about the thickness of a penny. Then, with a drill bit in a chuck mounted on the lathe tailstock, I bore a large starter hole in the wheel's center. The tip of an old file, ground like a regular lathe scraping tool, is used to enlarge the hole enough to slip-fit the bearings.

I use sealed ball bearings in both idler wheels, because they are durable and resist lateral movement of the wheels, which would lead to inaccurate tracking of the sanding belt. You can purchase the four ball bearings at most electric-motor repair shops and bearing distributors, or from W.W. Graingers (stock #1L050), 5959 W. Howard St., Chicago, Ill. 60648. These are the standard-size 203 sealed bearings used in almost all NEMA 56 frame-electric motors. The center shafts of old discarded motors (usually free of charge from motor-repair shops) are also a good source for the precision-ground stock needed to support the bearing. Many of these shafts have precision-ground bearing seats that will accommodate two bearings. Cut off this section. If the bearing seats are not long enough, see if the motor shop will press a couple of the shafts out of the rotors for you. To make dulling and mounting much easier, file, mill, sand or grind a flat on both sides of the portion of the shaft that will be resting on the machine. Don't flatten the actual seat.

Press the ball bearings on their shafts, coat the insides of the hardboard wheels with epoxy and insert the wheels over the bearings, being careful to leave them proud of the wheel. This will provide a surface for pressing the wheel on or off the shaft. Never put pressure against the hardboard wheels. Use sleeves to press against the inner edge of the bearings, if necessary, or you risk damaging the bearings. Let the wheels dry square and aligned with the shaft. Then, mount the drive wheel on the motor shaft and turn it true. If this wheel is ever removed, it may have to be turned true again.

The mandrel fits through the trunnion block and supports the laminated hardboard drive-shaft wheel. The trunnion block is routed and shaped to accept the trunnion cradle and worktable assembly.

The trunnion cradle is screwed to the table near the opening through which the belt runs. The bolt is needed to secure the cradle to the supporting trunnion block and to lock the table in place.

Trunnion assembly—The trunnion supporting the table is simply an arc of a circle that rocks in a round cradle around a common center point, as shown in the top photo at left. To ensure free and accurate movement, the center point of that circle should be right where the tabletop meets the front of the belt. The trunnion cradles are routed 5⁄16 in. deep in both sides of a 5½-in.-wide 8/4 upright with a circular router template. Leave the trunnion block about 2 ft. long to allow room for clamping the router template and to provide stock for the platen block, which supports the sanding belt. A disc is then lathe-turned to match the 5½-in. diameter of the circle cut by the router template. The disc is cut so the two equal halves plus twice the thickness of the table equals the disc's original diameter. You can make the template any way that's comfortable for you, but don't forget to account for the diameter of the router bushing you'll use so you can accurately produce recesses with the 2¾-in. radius shown in the drawing on the previous page. The template is also needed to rout a circle in a piece of scrap large enough to hold the trunnion discs when they're crosscut on the tablesaw. Also, rout another circle to serve as a diameter gauge when you turn the trunnion disc.

After turning the plywood trunnion disc to size, leave the piece on the faceplate and score a 5⁄16-in.-wide area ⅜ in. in from the outside rim. This area is used to lay out the slot that the trunnion clamping bolt runs in. It's a good idea to make extra circles in case you later want to make special grinding and sanding platforms for things such as lathe gouges and chisels. With your trunnion cut-off jig, cut the trunnion discs in equal parts, so when they are screwed on the bottom of the table, the tabletop will be at the center of the trunnion circle, as shown in the bottom photo at left.

Place the trunnion disc halves in their cradles so the tops of the trunnions are parallel to the top of the block; use a drill press to bore a ¼-in. hole through all three pieces. The hole should be positioned behind the center vertical line (about 1 in.) to give good clamping support. Take the trunnion halves out, mark where the bolt hole would be at the 45° setting, drill a hole through both pieces and cut out the slot on the jigsaw. Use a small rasp to do the final fitting, and elongate both ends of the slot slightly to allow for fine adjustments.

Now joint off about 1⁄16 in. from the top edge of the trunnion block so it will be slightly below the table's surface, and cut the block the same length as the base block. Bandsaw out an opening so the block will fit comfortably over the mandrel. Set a trunnion half in its cradle, pivot to the 45° setting, and mark and bandsaw a slope as shown on the front of the trunnion block to allow the table to drop.

Making the back post and top arm—Clamp the trunnion block in place over the base block, then clamp on the back post. Hold the top arm where you think it should go, giving it about a 10° droop, and clamp it in place for later trimming. Now make the two axle assemblies and clamp them in place. The top-wheel adjustment assembly shown in the top photo at right, is made from a piece of ¼x1½x6½-in. steel and a couple pieces of 1¼-in.-long, 1-in. by 1-in. angle iron. Drill and tap four 10-24 thread holes to secure the opposing pieces of angle iron. Drill two ³⁄₁₆-in. holes in each piece of angle iron, and slightly elongate the holes front to back for adjustment of the top-wheel shaft. Drill and tap the top-wheel axle with ¼-20 threads. The ⅝-in. diameter of the axle is a long way to thread, so a ¹³⁄₆₄-in. hole could be drilled about halfway down and the remainder tapped. Clamp the shaft between the angle-iron sides, and mark and drill two ¼-in. holes in the 6½-in. steel plate to mount the assembly to the arm. For final seating of the shaft in the top brackets, pinch the shaft with the top edges of the angle iron, and then tighten the angle-iron screws. Because the distance between the angle-iron sides is wider at the top than at the bottom, the shaft will seat quite snugly as the bottom adjustment screw pulls it down into the recess. Tracking adjustments are made by tightening and loosening the opposing thumbscrews.

Now, clamp the two axle-and-wheel assemblies in place. The front edge of the top wheel should be directly over the front edge of the bottom drive wheel. Be sure there is enough overhang on the back of the top arm to fasten the spring eyebolt. The back wheel should be high enough on the back post so it can be bolted on the front or back of the post, depending on whether you want to use 42-in. or 48-in. belts. For now, clamp it on the back, loop a 48-in. sanding belt over the wheels and move the arm around until it fits properly and secures the belt. Note the location of the top arm, and trim the back post accordingly.

Now take your hinge and determine the allowances needed for the knuckle, then cut the parts as shown in figure 1. Mount the hinge, and if everything fits okay, mount the top-wheel assembly. Take off the top arm, adjust the front bevel as needed so the arm can move freely and mount the eyebolts and the spring. (The spring I used came from a local hardware store's "Select-A-Spring" display, spring #171, for about $3.) Also at this time, predrill the back post for the screws that go through it into both the trunnion block and the base block. If you're sure of the position, you can also mount the back axle. In both cases of axle mounting, drill a hole in the wood about ¹⁄₃₂ in. larger than the bolt used, to allow for some adjustment.

With the sanding-belt system set for a 42-in. belt, make a template of the platen block. Get the remainder of the stock you used for the trunnion block and route a trunnion cradle in one side, a little deeper and a little farther back than the actual trunnion, as shown in the bottom photo at right, to allow free movement of the trunnions after the platen block is installed. Cut the platen block to length, allowing for the ⅛-in.-thick steel platen, and bandsaw to the needed profile. Install the platen, clamp the assembly in place and test-run a belt. The platen should deflect the belt out of line about ⅛ in. Unclamp the base and trunnion blocks and drill for platen-block mounting bolts.

Table assembly–Clamp the trunnion block to the base and clamp on the trunnion halves. Position the table, then clamp it to the trunnions. I let the table's centerline go right over one of the trunnions. Mark where the trunnions go, remove the table and drill countersunk holes in the table so it can be screwed to the trunnions. Reclamp the table to the trunnions and drill pilot holes in the trunnion halves, using the screw holes as guides. Drive the screws home while everything is still clamped. Next, mount the platen block and determine how much the table must be notched to fit over it. Allow for an extra ¼-in. clearance to the left of the platen

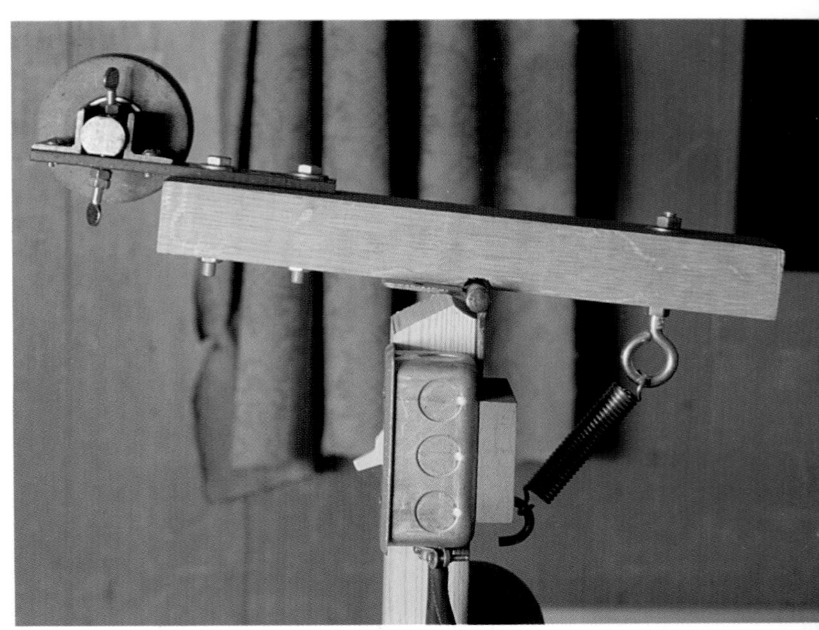

Two counteracting thumbscrews move the top-wheel shaft to control belt tracking. Belt tension is provided by the spring pulling against the hinged arm. A motor switch mounted on the same post is easy to reach from the front of the sander.

The platen block must be relieved so the trunnion cradles can be moved freely to adjust the angle of the sanding table.

for slipping belts in and out. Turn the table upside down and bandsaw out the clearance for the platen and block. Once the table and belt run satisfactorily, take the table off, turn it upside down and chisel out the relief for the 45° tilt.

Now unclamp everything and unscrew the base block from the plywood baseplate. Clamp the base and trunnion blocks in their proper position and install the screws through the back post into these two pieces, keeping everything as square as possible. Cut and bore a piece of flat strap steel, and screw it into the front of the trunnion block and into the base block. Reinstall the assembly on the plywood baseplate.

One point of caution: If a belt breaks, the top arm will flip upward about 4 in. in a sudden and unsettling manner. Keep away from the top wheel when sanding, to avoid a face full of sander. The spring does limit the upward movement of the top wheel, but a safety chain from the top arm to the front of the back post could easily be installed. You should also make hoods to cover the belts, pulley and wheels. A dust mask and eye protection are also recommended. ☐

Robert Vaughan is a professional woodworker and has his own shop in Roanoke, Va.

Bullnose Edge Sander

A low-cost method for sanding in tight places

by Lynn McSpadden

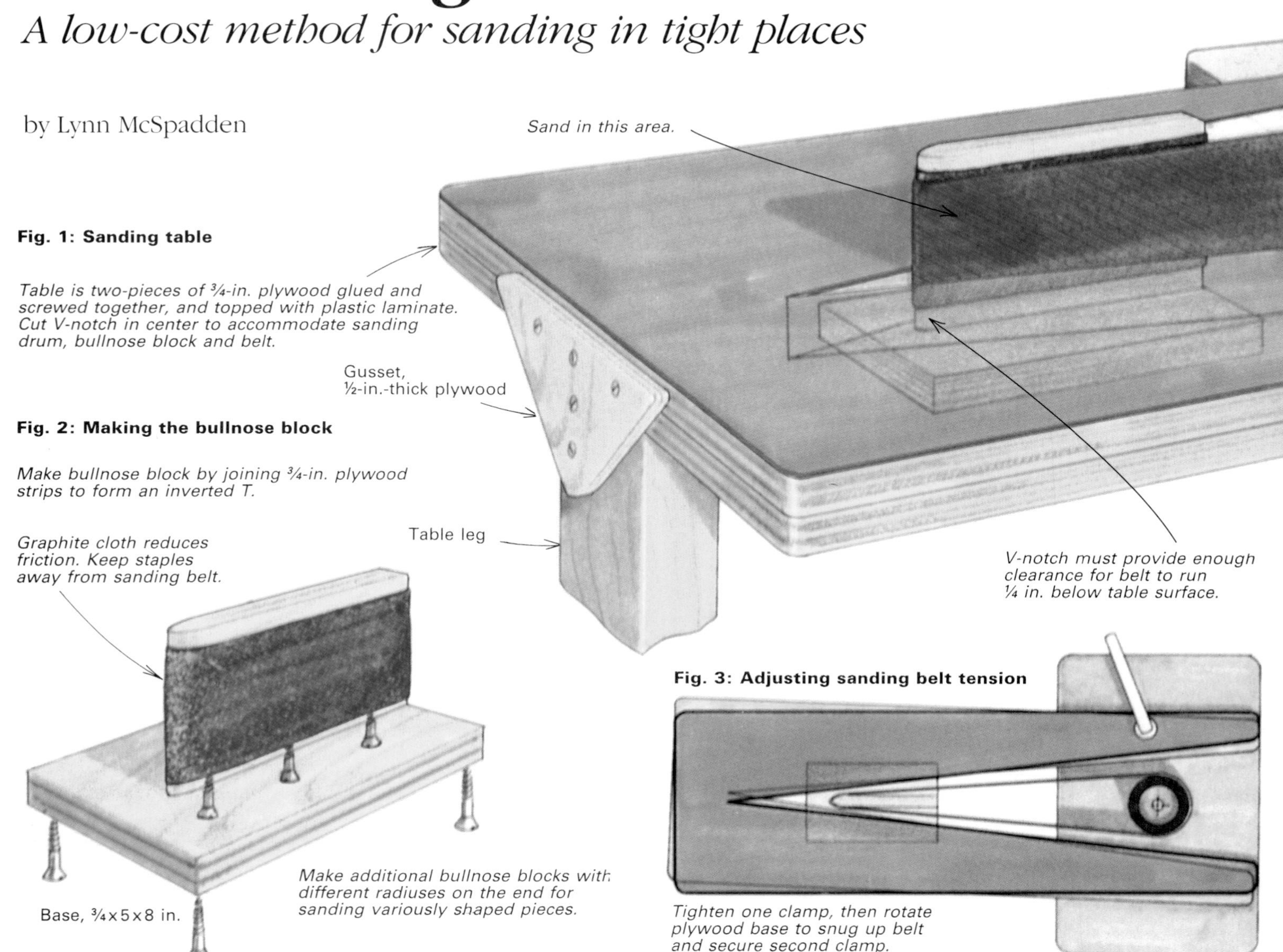

My company builds more than 1,500 mountain dulcimers a year. Until recently, we hand-sanded the scroll on every peg box. Since hand sanding was as boring as it was time consuming, I decided there had to be a power sander that could do the job more efficiently.

I first tried an oscillating spindle sander, but it proved to be just as slow as hand sanding and it left a bumpy surface. The edge belt sander I purchased operated at 3,030 sanding feet per minute (SFPM), and the belt tended to cut right through our delicate dulcimer parts. Flexible brush-style sanding wheels didn't work either, because they rounded edges too much.

Since none of the commercial sanding machines could end the boredom of hand sanding without sacrificing quality, I finally decided to build my own power sander. It cost me about $10 and is nothing more than a sanding belt that travels around a "bullnose" block, a radiused and tapered plywood strip that's mounted upright through the crotch of a deeply V-notched plywood base. The plywood base clamps to a drill-press table. The sanding belt is driven by a 3-in.-dia. rubber sanding drum chucked into the drill press. The belt rides just below the surface so the pieces will sand flush to the table.

I use the radiused end of the bullnose block to sand inside curves and its tapered sides to sand flat places. Adjusting the drill-press spindle RPM allows me to sand at 800 SFPM to 1,000 SFPM, a speed range that's slow enough to control material removal and slow enough to avoid burning the wood. Because the bullnose block is held in place with four wood screws, it is easily removable, making it simple to install differently radiused bullnose blocks to sand differently shaped workpieces.

Building the sanding table—I glued and screwed two ¾-in. pieces of plywood together to make a 1½ x 10 x 32-in. table. The dimensions depend on the size of your drill-press table and the length of your sanding belt. I V-notched the table to a length that would accommodate a 48-in. sanding belt, and I made the mouth of the V wide enough to accommodate a 3-in. sanding drum.

I found that clamping the long sanding table to the drill press caused the drill press to tip easily while sanding, so I mounted a leg to the front of the table. In addition to adding support, the leg also reduces vibration while sanding. Simply screw the leg to the end of the sanding table with a gusset made from ½-in. plywood (see figure 1).

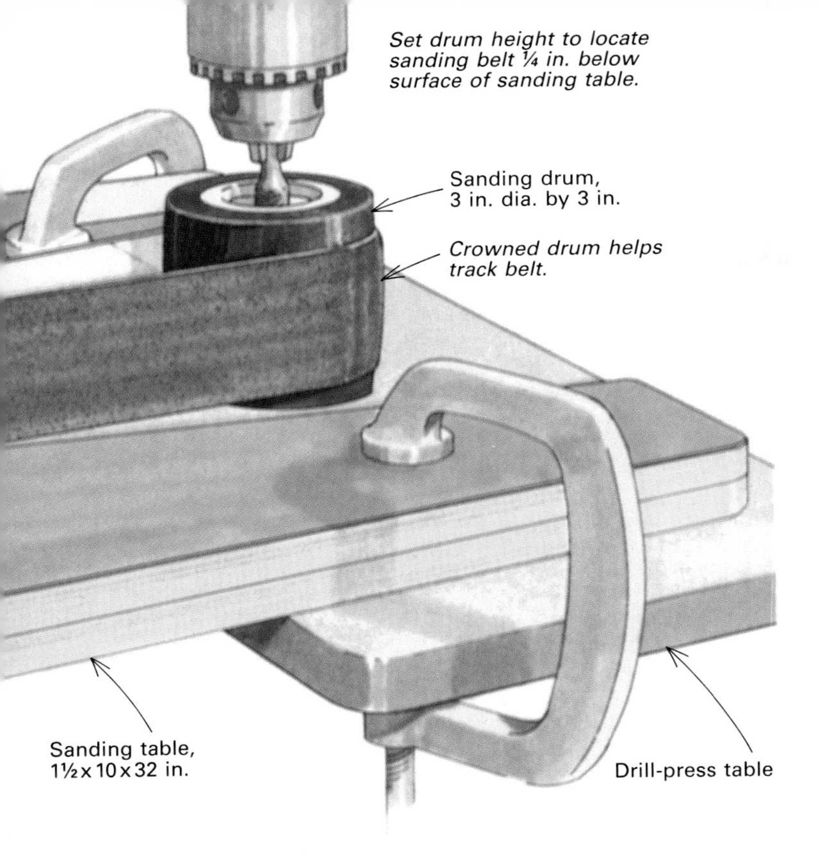

Because I wanted the pieces to slide easily, I cemented plastic laminate on the sanding table's top. A laminate trimmer or router with an edging bit works well for trimming the laminate to size and for notching it to match the base. So the bullnose block will seat properly and there will be adequate clearance for the sanding belt, file the crotch of the V in the laminate to a point that matches the crotch in the plywood base.

Making and mounting the bullnose–The base of the bullnose is made from ¾ x 5 x 8-in. plywood. The bullnose block is made from ¾-in. plywood, and it should be 8 in. long and tall enough to accommodate the width of the sanding belt. Radius the bullnose block with a belt sander, making sure the block remains square to the top and bottom. Sand the sides of the bullnose so the block tapers smoothly from the wide back edge to the radius in the front. This radius is determined by the size of the workpieces to be sanded. My bullnose block has a ¼-in. radius that fits the inside curves of the bandsawn dulcimer scrolls. Generally, the radius should be slightly smaller than the inside curve or recess to be sanded so the workpiece won't wedge against the belt. A smaller radius also gives more control, because the workpiece can be moved around on the bullnose block.

Glue and screw the bullnose block at 90° to the base and let it dry. The belt won't track properly unless the bullnose block is mounted at exactly 90°, front to back and side to side, so check both planes with a try square. Before using the bullnose, staple graphite cloth around the bullnose block (available from many supply houses, such as Derda Inc., 1195 E. Bertrand Road, Niles, Mich. 49120; 616-683-6666). This graphite cloth reduces friction between the back of the sanding belt and the edge and sides of the bullnose block. Locate the staples at the back of the bullnose block so they won't contact the sanding belt (see figure 2).

To ensure pieces are sanded flush to the table, the belt must travel at least ¼ in. below the sanding table's surface. This means the bullnose block must be mounted so the sanding belt can easily pass between the sides and crotch of the V and the radiused end and sides of the bullnose block. Push the bullnose block through the bottom of the table, with its radius facing the V crotch, so it protrudes through the sanding table's top. Then, back the bullnose block away from the crotch until there is enough clearance for the sanding belt, and secure it in place.

Setting up the sander–To mount the sander, loosely fasten the plywood base to the drill press with a couple of clamps. On a radial drill press (one with a tilting head) or on a drill press with a tilting table, the spindle must be squared to the table or the belt won't track correctly. Install a drill in the chuck and use a small try square to square the bit to the drill-press head or table.

Mount the rubber sanding drum in the drill-press spindle. A 3-in.-dia. by 3-in.-high sanding drum is available from Garrett Wade, 161 Ave. of the Americas, New York, N.Y. 10013; (800) 221-2942, or (212) 807-1757 in New York, Arkansas and Hawaii. Normally, rubber sanding drums are used with a sanding sleeve. A nut or screw on one end of the drum is tightened enough for the drum to expand and hold the sleeve in place. Instead of using a sanding sleeve on the drum, I overtighten the drum until it crowns in the center. Crowning increases the drum's center diameter, creating more tension in the belt's center, which helps the belt track straight. Install the sanding belt and raise or lower the quill until the belt is at least ¼ in. below the surface of the sanding table and the bottom edge of the belt is parallel with the entire length of the V in the sanding table.

After the belt is aligned, tighten one of the clamps, then rotate the plywood base until the belt tension is just tight enough so the sanding belt won't slip on the drum when the drill press is turned on (figure 3). If the tension is too high, the belt won't track properly, the sanded surface will be rough and the graphite cloth will wear out more quickly. Properly adjusted, the belt should run smooth and true and only slip when you apply excessive sanding pressure.

I used to make my own 48-in. sanding belts out of J-weight, 120-grit garnet paper, because its backing was strong yet flexible enough to travel around the tightly radiused bullnose block. (For more information on making sanding belts, see *FWW on Making and Modifying Machines,* p. 51.) Now I find it simpler to rip up standard J-weight, aluminum-oxide belts to the width I want. Using a sanding belt with the narrowest width possible for the work you're sanding will reduce friction at the bullnose block and help the belt track better.

It might be possible to use a shorter belt, but a wider bullnose block would be required to fit a wider and shorter V notch. A longer belt means a longer table must be made, and I don't think a longer belt would sand any better than a shorter one.

Adjusting sanding speed–The slower the sanding speed, the more control you'll have and the smoother the finish. The circumference of a 3-in.-dia. sanding drum is about 9½ in. So, with the spindle running at 1,000 RPM, the sanding belt is traveling 9,500 in. per minute. Dividing by 12 gives you 791 SFPM. Raise or lower the spindle RPM to adjust the belt to the SFPM you need.

When my company's production is in high gear, we run our bullnose sander eight hours a day for four consecutive days. The graphite cloth must be replaced every three to four hours. We go through about six sanding belts, but we make short work of the 1,500 peg boxes we used to hand-sand.

There was a time when I planned on making a much fancier model, complete with inlays and a couple of racing stripes to match our drill press. But, this first bullnose sander has worked so well from the start, I've decided to stay with it. □

Lynn McSpadden builds dulcimers and dulcimer kits. He lives in Mountain View, Ark.

From *Fine Woodworking* (July 1988) 71:76-77

Shopbuilt Thickness Sander

A low-cost alternative to handplaning

by S.R. Cook

Having earned my living from one form of woodworking or another for the last 15 years, I feel qualified to talk about ways of making sawdust. I began as an apprentice pipe organ and harpsichord maker, moved on to furniture, folk instruments, and cabinets, and, during the lean times, turned to plain old nail swatting. My style of woodwork, whether for cabinets or furniture, leans heavily toward frame-and-panel construction, using ¼-in.-thick solid wood panels rather than plywood. In the old days, I cut mortise-and-tenon joints for frames and handplaned each panel to its final thickness. Three kids and no savings account later, however, I began doweling all my cabinet joints and traded my meditative stints with handplanes for ear plugs, dust mask and belt sander. As I spent more and more time hanging on to that digging-in, corner-dipping belt sander, I yearned for a better way to surface wood. This yearning became a necessity when I fell for a $300 bargain and ended up with 1,500 bd. ft. of roughsawn birch from a local mill. After doing a little research on surfacing machines, I concluded that a power-feed drum sander was what I needed. The price was a bit of a snag, so I decided to build my own sander. My design was inspired by the planers offered as kits and plans by Kuster Woodworkers, P.O. Box 34, Skillman, N.J. 08558 (see box). A machinist friend and I modified the original idea to suit my needs and budget, and produced the machine shown on the facing page. It can sand panels up to 24 in. wide, down to 180 grit. With 36-grit abrasive, I can quickly dress a whole batch of rough lumber to a consistent thickness, then switch to finer grits and bring the lot to a smooth finish—all at a cost of $150 for parts and 50 hours assembly time.

My sander consists of three basic mechanical units: the sanding drum, the feed roller/speed reduction mechanism, and an extremely accurate table-height adjustment mechanism based on bicycle chains and sprockets. These parts are supported by a wooden box-like frame: the upper part of the box holds the drum and feed rollers, the lower helps support the table-height adjustment mechanism. Four sturdily braced legs attached to the box complete the machine.

The box must be strong and stable; I originally used 2-in. birch lumber, as shown, but later replaced this with 5½-in. by 24-in. sides made of two sheets of ¾-in. Baltic birch plywood, laminated face to face. Lay out the sides carefully, making sure both sides are square and mirror images of each other. Inaccuracies now will mean alignment problems later. After cutting the box pieces to size, glue the hardwood crosspieces and vertical supports to the side pieces. Reinforce the rabbet joints with long sheetrock screws, but don't add the screws until you've installed all the hardware, to make sure that the screws won't interfere with any mountings. For added rigidity, you might want to add a ¼-in. plywood bottom to the box. A 1-in.-square batten, glued and screwed to each end, accommodates the ends of the threaded adjustment rods.

When the box was completed, I added four legs long enough to raise the plywood sander table 34 in. off the floor. After using the sander, I decided it would be better to set the table at 30 in. to 32 in., about the height of my hands when they hang by my sides, to make it easier to lift the stock and feed it into the sander.

The table mechanism consists of four ½-in. threaded rods, one in each corner of the frame, as shown in the drawing. I had a machinist turn both ends of each rod down to a straight 5⁄16-in. shaft, so the end resembled the tenon and shoulder on a chair rung. The lower end of each rod sits in a 5⁄16-in. hole drilled in the corner of the frame's 1-in. by 1-in. lip, as shown on the facing page, far right, and its shoulders bear on a large washer embedded in the lip. The washer prevents the rod from wearing through the wood. The top end of each rod fits in an upper angle-iron support screwed to the frame.

Between the upper and lower supports, each pair of rods is threaded through a 28-in.-long piece of 1¼-in. angle iron, which is, in turn, screwed to a 1½-in.-thick laminated plywood table. The four sprockets bolted to the threaded rods are connected together with a taut length of roller chain (bicycle parts work well and are readily available). When you turn the adjustment wheel welded to the top of one of the front rods, all the threaded rods turn simultaneously, and the threading action raises or lowers the plywood table, giving you the ability to set the sander's depth of cut. I got my wheel from a scrapped tablesaw, but you could make one by brazing a metal handle to a steel disc.

To make the mechanism, bore a hole through the angle iron and weld a ½-in. nut over the hole. Then, thread a sprocket welded to a nut, a free nut, and one end of the 28-in. angle iron onto each rod. The free nut is used to lock the sprocket to the threaded rod. Don't install the chain until after the drum is aligned.

I made the sanding drum from a 24-in. length of 6-in. steel pipe. A machinist cut a lip inside the pipe to accept 5¾-in.-dia. discs cut from ¼-in. steel plate, and welded them in. Next, I bored a ¾-in. hole through the center of each end and ran a 32-in. by ¾-in. shaft down the length of the drum through the end caps, offsetting the shaft so it's longer on the drive-pulley side. After welding the shaft to the end caps, we chucked the entire assembly in a metalworking lathe and turned it true. The drum will probably still be out of balance and spin roughly. To check the balance, I slid the ball bearings onto the shaft, set the drum in the sander frame and spun it several times. If the drum always stops with the same side down, you know it's out of balance and the down side is

From *Fine Woodworking* (May 1986) 58:54-57

Steve Cook's shop-built abrasive sander, left, can flatten and thickness 24-in.-wide panels with 36-grit paper, then, with progressively finer grits, bring the whole batch to a smooth finish. Cook built the unit using bicycle parts, pipe, wood and commercially available rollers for about $150 and 50 hours assembly time. To mount the feed rollers, he slipped a copper pipe bushing over each shaft, added washers as shims to keep the roller from sliding back and forth, then secured the assembly with the spring-loaded wooden cap screws shown above. The sander table is adjusted by means of four threaded rods, right, running through nuts welded to an angle-iron frame. The bicycle chain connecting the sprockets on the rods makes it possible to raise or lower all four corners of the table simultaneously.

the heavier side. I corrected the imbalance by drilling shallow ¼-in. holes straight into the heavy side. Remove a little metal each time, and don't go all the way through the drum wall. I made about 50 holes before the heavier side seemed to disappear and the drum began to spin smoothly. Also, have the machinist mill a 2-in. start groove through one end of the drum to anchor sandpaper strips.

Next, install the bearings and wooden pillow blocks to hold the drum. I used caged automotive ball bearings with an inside diameter that fit over the ¾-in. shaft and a 2-in. outside diameter to fit the pillow blocks. On the drive side, bearings with double ball rows were used to accommodate side thrust. Slide the bearings over the ends of the drum shaft and position the shaft's long end on the drive-belt side. Place the bearing and shaft in the V-notches cut in the frame sides, place the notched caps over the bearings and bolt them down tightly. The pressure of the V-blocks is the only thing holding the bearings in place. To prevent the drum from moving left to right, I shimmed the space between the end of the drum and the sander frame with several large washers. The pulley or sprocket on each end of the shaft outside the frame secures those ends.

After installing the drum, I used the threaded rods to align the table parallel with the drum. Adjust the front side of the sander first. Turn the threaded rods individually to bring the table close to the drum. Then, using a long piece of ⅟₁₆-in. stock like a feeler gauge between the table and drum, twist one or both of the threaded rods until the wood gauge fits snugly along the whole length of the drum. Lock the sprocket in place by tightening the free nut against it. Remove the wooden gauge without moving the threaded rods, and repeat the adjustment process for the back side of the sander. Your table and drum should now be perfectly aligned. To make sure you don't lose this accurate adjustment, install the bicycle chain as tightly as possible. There shouldn't be any play in the chain or between the sprockets. The lengths sold by hardware stores and bicycle shops come with two master links, which are a snap to use for joining lengths of chain together.

Since the feed rollers are the most expensive parts of the sander, I tried to come up with a way of making them in the shop, but I've found no substitute for the commercially available models featuring a steel shaft bonded with a thick cushion of rubber. I ended up investing $76 for two rollers from Kuster Woodworkers. The rollers must be mounted keeping three things in mind: they turn at around 50 RPM, they have up and down movement of nearly ¼ in. (they should hang ⅛ in. to ³⁄₁₆ in. lower than the bottom of the drum for positive contact with the wood), and they must be fitted with stiff 1-in.-long, ½-in.-dia. coil springs to keep steady pressure on the wood being sanded. Each spring in the bushing blocks should exert about 20 lb. to 25 lb. of pressure. I used a short length of copper tubing as a bushing on either end of the roller, as shown above. The spring bears against the tube, which slides in the vertical 1½-in.-square notches cut in the frame sides. Pack each

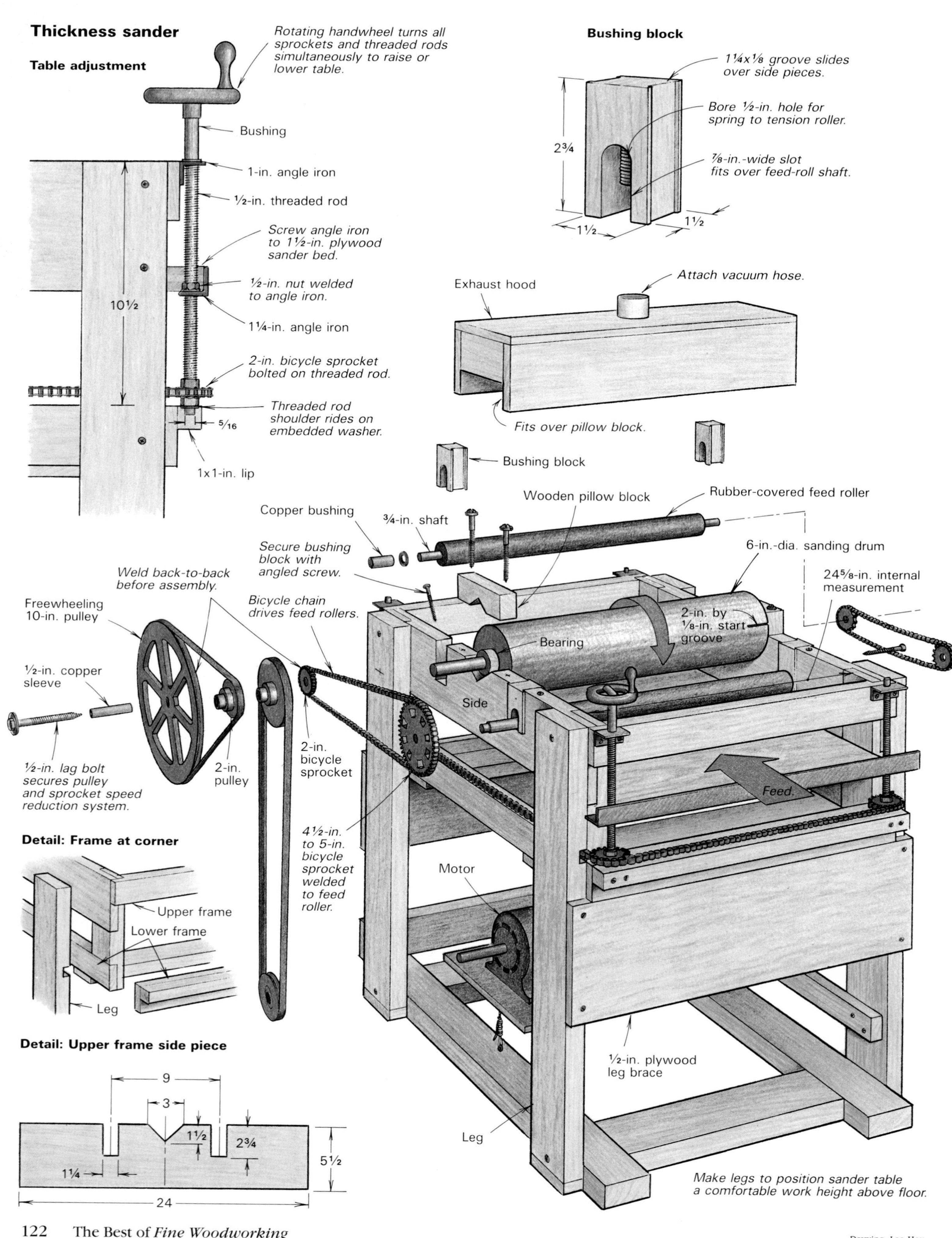

Drawing: Lee Hov

copper tube with grease before inserting the feed roller.

To drive the drum at about 1,200 RPM, I mounted a 3-in. pulley on the shaft of my 1,725-RPM, 1-HP electric motor. The drum rotates clockwise, in the same direction you're feeding in the wood, so dust builds up on the outfeed side and is carried away. My feed-roller drive consists of a 2-in. pulley on the drum shaft, driving a freewheeling 10-in. pulley and a 2-in. bike sprocket screwed into the frame. I welded the 10-in. pulley and 2-in. sprocket together and bored the unit to accept a piece of ½-in. copper pipe as a bushing, greased the inside, and mounted the unit to the frame with a ½-in.-thick lag bolt. You need that heavy lag bolt because it has to handle a great deal of torque here, due to the difference in diameters of the two pulleys. The freewheeling sprocket, in turn, drives via a bike chain, a 4½-in. or 5-in. sprocket welded on the end of the infeed roller. This arrangement produces a feed rate of 21 ft. per minute. On the other side of the machine, weld a 2-in. sprocket on each feed roller and connect them with a taut length of bike chain. This drives the rollers together. I line up the pulleys and sprockets by eye, sliding them on the shafts until they are aligned, then tighten the set screws or tack weld them in place.

I buy 3-in.-wide rolls of open-coat aluminum oxide paper that are 75-ft. or 150-ft. long (available from Kuster Woodworkers). Wider belts work too, but they're harder to put on. It takes about 12 ft. of 3-in. paper for the 24-in. drum, but for narrow stock it's not necessary to paper the entire drum. The best way I've found to attach the strips is to spray the drum with a light film of Weldwood Spray Glue adhesive available from local hardware stores, then immediately apply the sandpaper. I tape the end of the paper to fit the start groove, secure the end with a wooden shim, and wrap the paper on in a spiral fashion, as shown above right, in the direction opposite to the direction of drum rotation. Grit changes can be done in less than five minutes, and the paper stays put.

To operate the sander, put a rough board on the sander table and crank it up until the drum starts cutting. The maximum depth of cut with 36-grit paper is 1/32 in. If the feed jams during a cut, crank the table down and take a lighter cut. I use 36-grit for roughing stock to thickness, then progress to 80, 120, and finally, 180 grit. As each board comes out of the sander, whack it to remove some of the sawdust and continue planing. Keep the paper

To change sandpaper, Cook sprays adhesive on the sanding drum (note the holes drilled to balance the drum), then wraps on the abrasive. The paper spiral runs in the direction opposite to the drum rotation.

clean with a rubber sanding-belt cleaner. On the last pass, run each board through the sander twice without changing the depth setting. This will compensate for any table flexing and ensure that the stock is accurately flattened.

Except for the sandpaper changes, the sander doesn't require much maintenance. Keep the bushings greased. You might want to drill and tap the ends of the feed rollers and lag bolt for grease fittings and bore holes through the diameter of the shafts for grease flow. Unless you do all your work outside, you should also build a hood to go over the drum, so the machine can be hooked up to your shop vacuum or dust collection system. Otherwise, you'll have problems preventing the sawdust from clogging the machine, and your shop. The simple hood I made is shown in the drawing. Building a guard over the feed drive mechanism would be a good idea, too. Feel free to use your own ingenuity to improve on, or change, my basic sander. □

Steven R. Cook operates Pacific Rim Woodworking and Acoustic Keyboard Service in Edmonds, Wash.

An abrasive solution

by Curtis Erpelding

I'm a proponent of the hand-planed finish for one-of-a-kind pieces. Planing can be faster than sanding, and nothing can beat a hand-planed surface for clarity of figure and finish quality. But, for production work, handplanes can't always meet the demands of time and efficiency. I also hire assistants for production work, and it's not practical to teach part-time novice help to plane. For these reasons, I began investigating thickness sanders.

Thickness sanders looked more useful than belt sanders or stroke sanders. For my bent-laminated chairs, I need to surface 1/16-in. face veneers before gluing them to core laminates in forms. Prefinishing these faces eliminates the tedious job of sanding the curved surfaces after glueup. My other production work involves surfacing many dimensioned pieces, such as shelves, slats, box parts, and small panels. Belt sanding these parts wasn't faster than planing, and neither a belt sander nor a stroke sander was the solution for the veneer, even if I could handle the dust from a stroke sander. A thickness sander (I hoped) would handle the veneer, could quickly sand several pieces at once, and, with casters and port for a shop vacuum, fit efficiently into my work space.

I have friends who've had excellent results from simple hand-feed sanders, but I felt power feed was a must for production. Large abrasive-belt machines were out of my price range and even the Ultrasand, a ready-to-go drum sander manufactured by Kuster Woodworkers, P.O. Box 34, Skillman, N.J. 08558, was too large an investment for something I wasn't sure would work for me. After some deliberation, I chose the Kuster 24-in. Dynasand, a kit, which I thought I could adapt to suit my own needs.

I paid $660 for my 24-in. model, which included all metal parts and hardware, the gear motor to drive the feed rollers, and plans for a wooden base. I bought the wood and a 2-HP motor to drive the drum. I could have scrounged the parts more cheaply myself, but I'm glad I bought the kit and avoided a frustrating hunt for parts

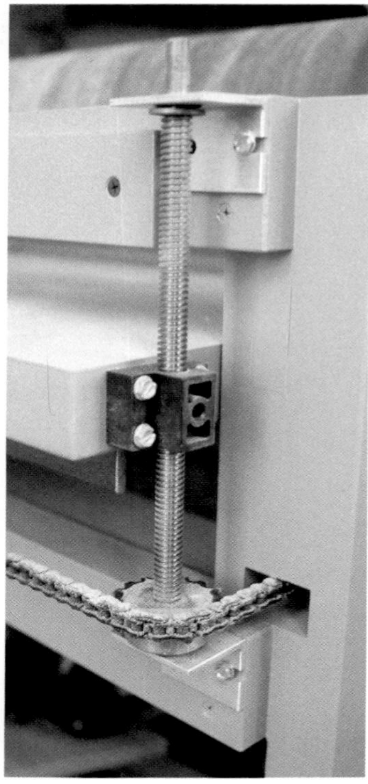

Curtis Erpelding built his 24-in. power-feed abrasive sander from a $660 kit, which he customized to handle the specialized needs of his production work. To eliminate sniping, he substituted angle-iron brackets for the machine's original aluminum brackets, which flexed enough to distort the critical alignment between the table, feed rollers and sanding drum.

and hardware. I could have designed my own machine, but the trial-and-error involved would have cost me more in time than it was worth. I wanted a proven design and the simple and elegant Kuster machine met the test.

Construction offered no real problems, the plans were clear and well ordered. Although I would have preferred stronger mortises and tenons, I stuck to the half-lap joints in the plans, rather than risk altering dimensions and structure, and possibly creating assembly problems. The company had obviously invested a lot of time in working the bugs out of the design.

Some of the Kuster parts were crudely machined, though not to the detriment of function. The pillow-block bearings, however, were first rate, and the 6-in. aluminum drum was a thing of beauty, lightweight and true. I painted the frame with industrial floor enamel to give the machine a professional look and make it easier to clean. Electrical cord, plugs, switches and adapters ran about $30.

The completed machine performed famously. It easily sanded out the knife marks, pits and ridges characteristic of thick-sliced veneer. It even sanded badly cupped veneer, the feed rollers gently flattening the wood and easing it past the sanding head. I did add a metal baffle on the outfeed side to keep the sprung-back veneer from hanging up on the frame.

It worked as well surfacing thicker pieces, eliminating planer marks in one pass with 180 grit. But, it did create two problems which, luckily, were solvable. First, though it removed the planer marks, it left its own faint chatter marks. Secondly, it "sniped" the boards, leaving noticeable ridges on both ends of each board.

Chatter marks occur because even the slightest out-of-roundness or runout in the drum or bearings creates a definite washboard drum action. The slower the feed rate, the less noticeable the problem. Since the standard feed rate is already a snail-paced 12 ft. per minute, I needed another solution. By wrapping felt around the drum underneath the abrasive, I dampened out the marks almost entirely. For the finest work I then wet the grain, and, when dry, used a vibrating sander or hand block to remove any residual chatter and scratch marks.

The problem of snipe cannot be explained without noting one feature that surprised me—the feed rollers and the sanding drum rotate in the same direction, unlike a thickness planer, where the feed rollers and cutterhead rotate in opposite directions. After I tensioned the feed rollers as much as possible, thinking that necessary to hold the work, I found the machine sniped about 4 in. from each end of the board. The .007-in.-deep mark was quite visible. I called the factory and Bob Kuster (I found the folks at Kuster friendly and helpful) suggested I shim the bottom of the feed rollers up to just 1/16 in. below the sanding drum and put only as much tension on them as necessary to feed the work. These prescriptions reduced the depth of snipe about .005 in., leaving just the faintest ridge at the points where the sanding head is when the front edge of the board hits the outfeed roller and when the back edge of the board leaves the infeed roller. With a dial indicator, I traced the problem to the lower brackets that support the threaded rods which, in turn, support the table. The rather thin aluminum brackets deflected .002 in. as a board was fed through. Replacing these with thicker steel angle solved the problem. The critical factor apparently is the difference in distance from the table between the feed rollers and the sanding head. And, this is where I made another modification.

The absolute minimum depth of the feed rollers below the sanding drum is 1/16 in. Any less and the drum *will* catch the work and fling it through. Unfortunately, even something as simple as changing from 180-grit cloth to 36-grit cloth may reduce the distance beyond the minimum. For these adjustments, I put wood screws under the bushings that house the rollers. By turning these in or out I can lower or raise the rollers. Unfortunately it's hard to gauge the adjustment because the roller must be removed to get at the screws. A better solution would be machine screws through T-nuts mounted beneath the frame. This would be easy to install before the frame was assembled, but unfortunately, none of this is mentioned in the plans.

Even with proper tension and adjustment, the sanding drum can still catch and propel work out the outfeed side if you simultaneously feed in several pieces of different thicknesses. The feed rollers will center on the thicker pieces, allowing the drum to grab the thinner ones. The problem can also occur with uneven dust build-up on the outfeed roller. Multiple feeds must be the same thickness and the rollers must be kept clean. In any event, don't stand directly behind the machine when retrieving work coming through.

The hood I built for my sander accommodates my shop vacuum nozzle, but the vacuum catches only the fine airborne dust. The thick layer deposited on the board must be brushed off before sending the board through again, to prevent build-up on the rollers and resulting slippage.

I wanted a machine that would do surface sanding, sand more than one piece at once, be easy for a novice helper to operate, and fit into the scale and purpose of my shop. The Dynasand met or exceeded my expectations in every case. □

Curtis Erpelding is a woodworker and designer in Seattle, Wash.

Shop-Made Sanding Drums

Cylinders turned true without a lathe

by Tim Hanson

A few months after I'd completed kitchen cabinets for our home, my good wife spotted a Parsons bench she admired, so I agreed to make one. Now, a Parsons bench has a lot of bandsawn curves and piercework designs in the back, all of which require sanding. After plenty of hard, slow work with a belt sander, I realized that what I really needed were some good-sized sanding drums I could mount in my drill press. You can buy these drums, but they're never the right size, so I started tinkering and came up with a method to make my own.

For precise work, a sanding drum's circumference needs to be exactly concentric to the arbor upon which the drum is mounted. To achieve this concentricity, I first mounted the arbor into a block of wood, then devised a way to turn the block perfectly cylindrical on my drill press (this could, of course, be done on a lathe instead). Rather than gluing the paper onto the drum or bothering with a sleeve, I designed a way to wrap regular sandpaper around the drum so it can be pulled tight and fastened. This method worked so well that I made an entire set of drums, ranging in diameter from 1¼ in. to 3⅛ in. One drum has a handle opposite the arbor end, so I can chuck it into a portable drill.

Here's how I make the drums: For a 2½-in.-dia., 5½-in.-long drum (this size uses exactly half of a 9-in. by 11-in. sheet of sandpaper), you need a 3-in.-sq. blank of wood. I've used maple, oak and poplar, but any hardwood will do. With a tablesaw, cut off the blank's corners to speed turning later (see figure 1 on p. 126). Next, thread a 6-in.-long, ½-in.-dia. machine bolt into a 13⁄32-in. hole bored 3½ in. deep into the center of the blank. Ream the hole ½ in. in diameter to the depth of the unthreaded portion of the bolt. Tighten the bolt until it just bottoms out. Good and snug is tight enough. To keep from stripping the hole, mark the hole depth on the bolt so you can tell when it bottoms.

To true-up the cylinder, I needed some sort of a cutting tool to mount on the drill-press table, so I ground the head of a 1½-in. drywall screw to a beveled edge, then honed the edge on an oilstone. It may not be fine tool steel, but a drywall screw is case-hardened and sharpens fairly well. I screwed my cutter into a block of wood at about a 20° angle, leaving ⅛ in. to 3⁄16 in. projecting. I then clamped the block to my drill-press table, which rotates to the vertical position, easily allowing the cutting edge to contact the spinning drum. If your table doesn't rotate, swing it to one side and clamp the block at the edge of the table, then swing the table toward the spinning blank to adjust the depth of cut. With the drill-press turning at 900 RPM, I cut about 1⁄64 in. per pass by raising and lowering the quill. My drill press has only 3½ in. of quill travel, so on a 5½-in.-long cylinder, I had to adjust the table's height to finish turning. When you've turned the drum down, finish the cylinder with a flat sanding block. Finally, drill a hole through the side of the drum and into the bolt for a 6D finishing nail that serves as a locking pin.

A friend of mine encountered chatter when he tried turning drums on his drill press, which is a good bit older than mine and has looser bearings and quill. We solved this problem by screwing another drywall screw right in front of, and 3⁄32 in. below, the cutter screw. The second screwhead is a few thousandths of an inch below the cutter, and it seems to brace the cutter against chatter in the same way a chip breaker works in a handplane.

The sandpaper locking pin can be made in one of two ways. For large drums, I use the cam-type pin made from a ½-in. dowel, as shown in figure 3. You can lock the paper on a smaller drum with the simple angled end pin, also depicted in the drawing. To make a cam-type pin, cut a slot for a screwdriver into the end of a dowel section. A ⅝-in. copper-tubing ferrule driven onto the end of the dowel will keep it from splitting when the paper is tightened. Now, with a backsaw, make a straight cut about ⅛ in. deep down the length of the dowel and use a chisel to carve the cam shape shown in the drawing. The pin fits into a ½-in. hole bored down through the drum. Locate the hole so you can leave 3⁄16 in. of wood between the drum's outer edge and the edge of the hole. Cut a slot for the sandpaper by angling a backsaw cut from the outside of the drum to the tangent of the ½-in. hole. Clean the slot up with sandpaper so it's 3⁄32 in. wide. To install sandpaper, cut a strip to length and tuck both ends into the slot, then insert the pin and turn it counterclockwise with a screwdriver to snug up the paper.

On small drums with no room for a ½-in. hole, use angled locking pins. For these pins, the hole must be about 1⁄16 in. larger than the dowel diameter, but the slot size is the same. Insert both ends of the paper, then use the angled end of the pin to crimp the paper as you slowly work the pin into the hole, turning it back and forth as you go. For paper changes, tap the pin out with a punch.

A sanding drum with a handle at one end is made the same way, except the shaft is pinned in the block before the cylinder is turned and a ⅜-in. steel rod rather than a bolt serves as the arbor. Set the nail well below the drum's surface so it won't foul your cutter. Figure 4 on p. 126 shows the details.

I operate the 3-in. drum at about 900 RPM. High speeds or lots of pressure burn the work or load up the paper. I use 40-grit for fast stock removal, 80-grit and 120-grit for finishing passes. That's another nice thing about making your own drums: You can choose any grit available in standard sandpaper sheets. □

Tim Hanson is a retired contractor and woodworker. He lives in Indianapolis, Ind.

From *Fine Woodworking* (November 1987) 67:68-69

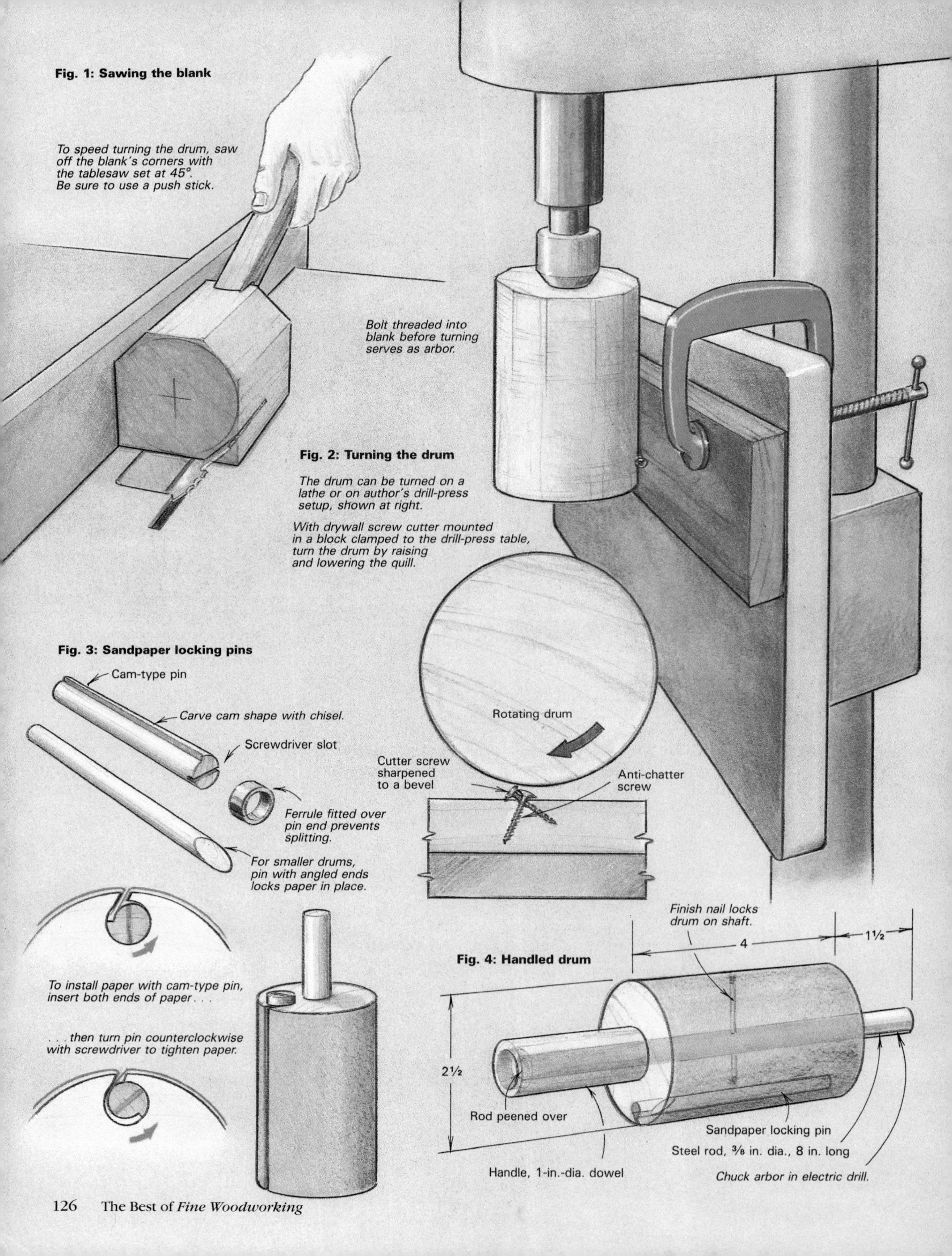
Fig. 1: Sawing the blank
To speed turning the drum, saw off the blank's corners with the tablesaw set at 45°. Be sure to use a push stick.
Bolt threaded into blank before turning serves as arbor.
Fig. 2: Turning the drum
The drum can be turned on a lathe or on author's drill-press setup, shown at right.
With drywall screw cutter mounted in a block clamped to the drill-press table, turn the drum by raising and lowering the quill.
Rotating drum
Cutter screw sharpened to a bevel
Anti-chatter screw
Fig. 3: Sandpaper locking pins
Cam-type pin
Carve cam shape with chisel.
Screwdriver slot
Ferrule fitted over pin end prevents splitting.
For smaller drums, pin with angled ends locks paper in place.
To install paper with cam-type pin, insert both ends of paper . . .
. . . then turn pin counterclockwise with screwdriver to tighten paper.
Finish nail locks drum on shaft.
4
1½
Fig. 4: Handled drum
2½
Rod peened over
Sandpaper locking pin
Steel rod, ⅜ in. dia., 8 in. long
Handle, 1-in.-dia. dowel
Chuck arbor in electric drill.

Index

O

P

R

S

T

V

W